ANIMAL TRACKS

of

NEVADA &
THE GREAT BASIN

Tamara Eder

with contributions from Ian Sheldon

© 2002 by Lone Pine Publishing
First printed in 2002 10 9 8 7 6 5 4 3 2 1
Printed in Canada

THE PUBLISHER: LONE PINE PUBLISHING

1901 Raymond Avenue SW, Suite C	10145-81 Avenue
Renton, WA 98055	Edmonton, AB T6E 1W9
USA	Canada

Lone Pine Publishing Website: http://www.lonepinepublishing.com

National Library of Canada Cataloguing in Publication Data

Eder, Tamara, (date)
 Animal tracks of Nevada and the Great Basin

 Includes bibliographical references and index.
 ISBN 1-55105-339-X

 1. Animal tracks—Great Basin—Identification. I. Sheldon, Ian, (date) II. Title.
QL768.E373 2001   591.47'9   C2001-911032-4

Editorial Director: Nancy Foulds
Editor: Volker Bodegom
Proofreaders: Lee Craig, Genevieve Boyer
Production Coordinator: Jen Fafard
Design, Layout & Production: Volker Bodegom, Monica Triska
Cover Design: Elliot Engley
Technical Contributor: Mark Elbroch
Cartography: Volker Bodegom
Animal Illustrations: Gary Ross, Horst Krause, Ian Sheldon,
 Ewa Pluciennik
Track Illustrations: Ian Sheldon
Scanning: Elite Lithographers Ltd.

We acknowledge the financial support of the Government of Canada through the Book Publishing Industry Development Program (BPIDP) for our publishing activities.

PC: P4

CONTENTS

INTRODUCTION

If you have ever spent time with an experienced tracker, or perhaps a veteran hunter, then you know just how much there is to learn about the subject of tracking and just how exciting the challenge of tracking animals can be. Maybe you think that tracking is no fun, because all you get to see is the animal's prints. What about the animal itself—is that not much more exciting? Well, for most of us who don't spend a great deal of time in the beautiful wilderness of the Great Basin states, the chances of seeing the the swift Pronghorn or the elusive Kit Fox are slim. The closest that we may ever get to some animals is through their tracks, but the tracks can still inspire a very intimate experience. Remember, you are following in the footsteps of the unseen—animals that are in pursuit of prey, or perhaps being pursued as prey.

This book offers an introduction to the complex world of tracking animals. Sometimes tracking is easy. At other times it is an incredible challenge that leaves you wondering just what animal made those unusual tracks. Take this book into the field with you, and it can provide some help with the first steps to identification. Animals tracks and trails are this book's focus; you will learn to recognize subtle differences for both. There are, of course, many additional signs to consider, such as scat and food caches, all of which help you to understand the animal that you are tracking.

Remember, it takes many years to become an expert tracker. Tracking is one of those skills that grows with you as you acquire new knowledge in new situations. Most importantly, you will have an intimate experience with nature. You will learn the secrets of the seldom-seen. The more you discover, the more you will want to know and, by developing a good understanding of tracking, you will gain an excellent appreciation of the intricacies and delights of our marvelous natural world.

How to Use This Book

Most importantly, take this book into the field with you! Relying on your memory is not an adequate way to identify tracks. Track identification has to be done in the field, or with detailed sketches and notes that you can take home.

Black-tailed Jackrabbit

Much of the process of identification involves circumstantial evidence, so you will have much more success when standing beside the track.

This book is laid out in an easy-to-use format. Beginning on p. 150, there is a quick reference appendix to the tracks of all the animals illustrated in the book. This appendix provides a fast way to familiarize yourself with certain tracks, and it guides you to the more informative descriptions of each animal and its tracks.

Each description is illustrated with the appropriate footprints and the track patterns that it usually leaves. Although these illustrations are not exhaustive, they do show the tracks or groups of prints that you will most likely see. You will find a list of dimensions for the tracks, giving the general range, but there will always be extremes, just as there are with people who have unusually small or large feet. Under the category 'Size' (of animal), the 'greater-than' sign (>) indicates a large size difference between the sexes.

If you think that you may have identified a track, check the 'Similar Species' section. This section is designed to help you confirm your conclusions by pointing out other animals that leave similar tracks and by showing you ways to distinguish among them.

As you read this book, you will notice an abundance of words such as 'often,' 'mostly' and 'usually.' Unfortunately, tracking will never be an exact science; we cannot expect animals to conform to our expectations, so be prepared for the unpredictable.

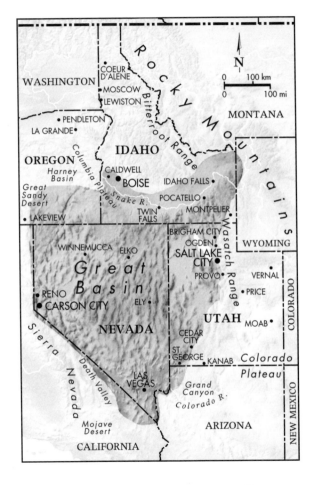

WASHINGTON

COEUR D'ALENE
MOSCOW
LEWISTON

PENDLETON
LA GRANDE

MONTANA

OREGON

Harney Basin

Great Sandy Desert

Columbia Plateau

IDAHO

CALDWELL
BOISE

Snake R.

IDAHO FALLS
POCATELLO
MONTPELIER

Bitterroot Range

ROCKY Mountains

LAKEVIEW

TWIN FALLS

WINNEMUCCA

ELKO

Great Basin

RENO
CARSON CITY

ELY

NEVADA

Sierra Nevada

Death Valley

LAS VEGAS

Mojave Desert

CALIFORNIA

BRIGHAM CITY
OGDEN
SALT LAKE CITY
PROVO

Wasatch Range

WYOMING

VERNAL
PRICE

UTAH

MOAB

COLORADO

CEDAR CITY
ST. GEORGE
KANAB

Colorado Plateau

Grand Canyon
Colorado R.

ARIZONA

NEW MEXICO

N

0   100 km
0   100 mi

Tips on Tracking

As you flip through this guide, you will notice clear, well-formed prints. Do not be deceived! It is a rare track that will ever show so clearly. For a good, clear print, the perfect conditions are slightly wet, shallow snow that is not melting, or slightly soft mud that is not actually wet. These conditions can be rare—most often you will be dealing with incomplete or faint prints, where you cannot even really be sure of the number of toes.

Should you find yourself looking at a clear print, then the job of identification is much easier. There are a number of key features to look for: measure the length and width of the print, count the number of toes, check for claw marks and note how far away they are from the body of the print, and look for a heel mark.

Bighorn Sheep

Keep in mind more subtle features, such as the spacing between the toes, whether or not they are parallel, and whether fur on the sole of the foot has made the print less clear.

When you are faced with the challenge of identifying an unclear print—or even if you think that you have made a successful identification from one print alone—look beyond the single footprint and search out others. Do not rely on the dimensions of just one print, but collect measurements from several prints to get an average impression. Even the prints within one trail can show a lot of variation.

Try to determine which is the fore print and which is the hind, and remember that many animals are built very differently from humans, having larger forefeet than hind feet. Sometimes the prints will overlap, but at other times they can be directly on top of one another in a direct register. For some animals, the fore and hind prints are pretty much the same.

Mountain Lion

Check out the pattern that the tracks make together in the trail, and follow the trail for as many paces as is necessary for you to become familiar with the pattern. Patterns are very important, and can be the distinguishing feature between different animals with otherwise similar tracks.

Follow the trail for some distance—it can give you some vital clues. For example, the trail may lead you to a tree, indicating that the animal is a climber, or it may lead down into a burrow. This part of tracking can be the most rewarding, because you are following the life of the animal as it hunts, runs, walks, jumps, feeds or tries to escape a predator.

Take into consideration the habitat. Sometimes habitat alone will allow you to distinguish very similar tracks—one species might be found on riverbanks, whereas another might be encountered just in dense forest.

*Greater
Roadrunner*

Think about your geographical location, too, because some animals have a limited range. This consideration can rule out some species and help you with your identification.

Remember that every animal will at some point leave a print or trail that looks just like the print or trail of a completely different animal!

Finally, keep in mind that if you track quietly, you might catch up with the maker of the prints.

Terms and Measurements

Some of the terms used in tracking can be rather confusing, and they often depend on personal interpretation. For example, what comes to your mind if you see the word 'hopping'? Perhaps you imagine a person hopping about on one leg, or perhaps you imagine a rabbit hopping through the countryside. Clearly, one person's perception of motion can be very different from another's. Some useful terms are explained on the next few pages to clarify what is meant in this book, and, where appropriate, how the measurements given fit in with each term.

The following terms are sometimes used loosely and interchangeably—for example, a rabbit might be described as a 'hopper' and a squirrel as a 'bounder,' yet both leave the same pattern of prints in the same sequence.

Ambling: Fast, rolling walking.

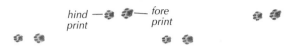

Bounding: A gait of four-legged animals in which the two hind feet land simultaneously, usually registering in front of the fore prints. It is common in rodents and the rabbit family. 'Hopping' or 'jumping' can often be substituted.

Gait: Describes how an animal is moving at some point in time. Different gaits result in different observable trail characteristics.

Galloping: A gait used by animals with four legs of even length, such as dogs, moving at high speed, hind feet registering in front of forefeet.

Hopping: Similar to bounding. With four-legged animals it is usually indicated by tight clusters of prints, fore prints set between and behind the hind prints. A bird hopping on two feet creates a series of paired tracks along its trail.

Loping: Like galloping but slower, with each foot falling independently and leaving a trail pattern that consists of groups of tracks in the sequence fore-hind-fore-hind, usually roughly in a line.

Mustelids (weasel family) often use ***2×2 loping***, in which the hind feet register directly on the fore prints. The resulting pattern has angled, paired tracks.

Running: Like galloping, but applied generally to animals moving at high speed. Also used for two-legged animals.

Stotting (applies to the Mule Deer only): Describes the action of taking off from the ground and landing on all four feet at once, in pogo-stick fashion.

Trotting: Faster than walking, slower than running. The diagonally opposite limbs move simultaneously; that is, the right forefoot with the left hind, then the left forefoot with the right hind. This gait is the natural one for canids (dog family), short-tailed shrews and voles.

hind
print

fore
print

Canids may use ***side-trotting***, a fast trotting in which the hind end of the animal shifts to one side. The resulting track pattern has paired tracks, with all the fore prints on one side and all the hind prints on the other.

Walking: A slow gait in which each foot moves independently of the others, resulting in an alternating track pattern. This gait is common for felines (cat family) and deer, as well as wide-bodied animals, such as bears and porcupines. The term is also used for two-legged animals.

Other Tracking Terms:

Dewclaws: Two small, toe-like structures set above and behind the main part of the foot of most hoofed animals.

Direct Register: The hind foot falls directly on the fore print.

double register   *direct register*

Double Register: The hind foot registers, overlapping the fore print only slightly or falling beside it, so that both prints can be seen at least in part.

Dragline: A line left in snow or mud by a foot or the tail dragging over the surface.

dragline

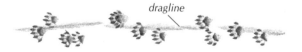

Gallop Group: A track pattern of four prints made at a gallop, usually with the hind feet registering in front of the forefeet (see '**galloping**' for illustration).

Height: Taken at the animal's shoulder.

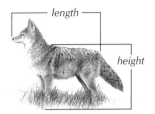

Length: The animal's body length from head to rump, not including the tail, unless otherwise indicated.

Metacarpal Pad: A small pad near the palm pad or between the palm pad and heel on the forefeet of bears and members of the weasel family.

Print (also called '**_track_**'): Fore and hind prints are treated individually. Print dimensions given are 'length' (including claws—maximum values may represent occasional heel register for some animals) and 'width.' A group of prints made by each of the animal's feet makes up a track pattern.

Register: To leave a mark—said about a foot, claw or other part of an animal's body.

Retractable: Describes claws that can be pulled in to keep them sharp, as with the cat family; these claws do not register in the prints. Foxes have semi-retractable claws.

Sitzmark: The mark left on the ground by an animal falling or jumping from a tree.

Straddle: The total width of the trail, all prints considered.

Stride: For consistency among different animals, the stride is taken as the distance from the center of one print (or print group) to the center of the next one. Some books may use the term 'pace.'

Track: Same as '***print***.'

Track Pattern: The pattern left after each foot registers once; a set of prints, such as a gallop group.

Trail: A series of track patterns; think of it as the path of the animal.

Kit Fox

MAMMALS

River Otter

Bison

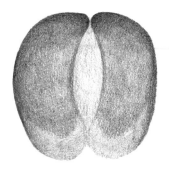

Fore and Hind Prints
Length: 4–6 in (10–15 cm)
Width: 4–6 in (10–15 cm)

Straddle
10–21 in (25–55 cm)

Stride
Walking: 14–32 in (35–80 cm)

Size (bull>cow)
Height: 5–6 ft (1.5–1.8 m)
Length: 10–12 ft (3–3.7 m)

Weight
Male: 800–2000 lb (360–900 kg)
Female: 700–1100 lb (320–500 kg)

walking

BISON
(Buffalo)
Bison bison

An estimated 70 million Bison once roamed North America. As few as 1500 survived the wholesale Bison slaughter of the nineteenth century. As a result of a major effort to save this magnificent beast from extinction, hundreds of thousands of Bison now inhabit scattered protected areas and ranches across the continent. All Bison in the Great Basin are confined by fences.

In a Bison's alternating walking pattern, the slightly smaller hind foot usually registers on or near the fore print. On firm ground, only the outer edge of the hoof may register, but in soft mud or snow the whole foot registers—perhaps the dewclaws too. Foot drag is common. The abundant 'pies' may be mistaken for those of domestic cattle. Additional signs of Bison include rubbing posts or trees that have tufts of distinctive brown hair hanging from them and the large pits in which Bison wallow. Do not be fooled by a Bison's calm exterior—a hefty bull Bison can inflict serious injury!

Similar Species: Domestic Cattle (*Bos* spp.) prints are similar. On firm surfaces, Horse (p. 34) prints can resemble Bison prints.

Elk

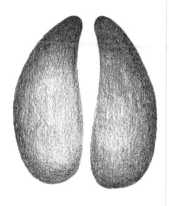

Fore and Hind Prints
Length: 3.2–5 in (8–13 cm)
Width: 2.5–4.5 in (6.5–11 cm)

Straddle
7–12 in (18–30 cm)

Stride
Walking: 16–34 in (40–85 cm)
Galloping: 3.3–8 ft (1–2.4 m)
Group length: to 6.3 ft (1.9 m)

Size (bull>cow)
Height: 4–5 ft (1.2–1.5 m)
Length: 6.5–10 ft (2–3 m)

Weight
500–1000 lb (225–450 kg)

gallop print | *walking*

ELK
(Wapiti)
Cervus elaphus

The Elk is common in the meadows and open forests of the mountains and foothills. Female Elk and young are often seen in social herds. They like to feed in forest openings and mountain meadows, moving into valleys when winter sets in. The Stag, which prefers to go solo, is easily recognized by his magnificent rack of antlers and distinctive bugling. A good place to look for Elk tracks is in the soft mud by summer ponds, where these animals like to drink and sometimes splash around.

Elk leave a neat alternating walking pattern of large, rounded prints, often in well-worn winter paths. The hind foot sometimes double registers slightly in front of the fore print. In deeper snow, or if an Elk gallops (with its toes spread wide), the dewclaws may register.

Similar Species: Deer (pp. 24–27) and Pronghorn Antelope (p. 28) prints can be difficult to distinguish, but they are generally smaller.

Mule Deer

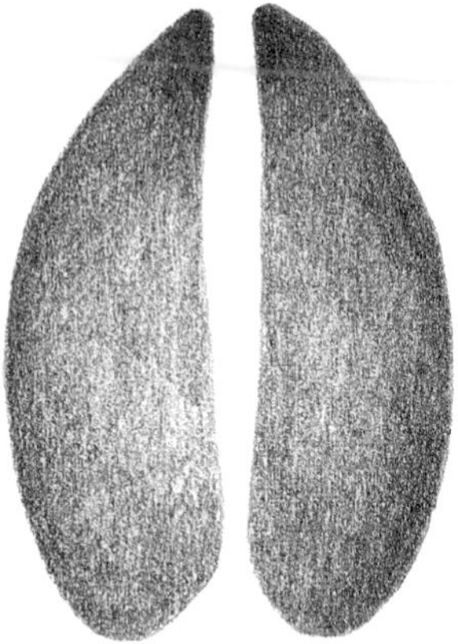

Fore and Hind Prints
Length: 2–3.3 in (5–8.5 cm)
Width: 1.6–2.5 in (4–6.5 cm)

Straddle
5–10 in (13–25 cm)

Stride
Walking: 10–24 in (25–60 cm)
Stotting: 9–19 ft (2.7–5.8 m)

Size (buck>doe)
Height: 3–3.5 ft (90–110 cm)
Length: 4–6.5 ft (1.2–2 m)

Weight
100–450 lb (45–200 kg)

walking 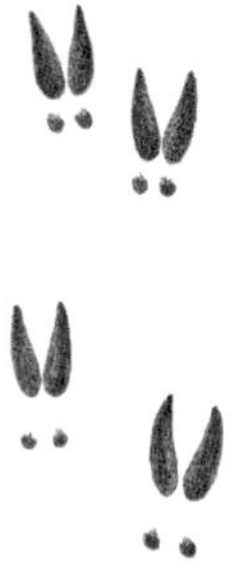   *stot group*

MULE DEER
(Black-tailed Deer)
Odocoileus hemionus

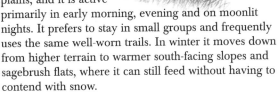

The widespread Mule Deer frequents meadows, open woodlands and plains, and it is active primarily in early morning, evening and on moonlit nights. It prefers to stay in small groups and frequently uses the same well-worn trails. In winter it moves down from higher terrain to warmer south-facing slopes and sagebrush flats, where it can still feed without having to contend with snow.

The Mule Deer's neat alternating walking pattern shows the hind foot registered on top of the fore print. In deep mud or snow, or when the animal is moving quickly, the dewclaws will register, closer to the hoof on the fore prints than on the hind ones. At high speed this deer has a unique gait—stotting—in which it jumps with all its feet leaving or striking the ground at once. The stotting track pattern shows how the toes splay to distribute the weight and give better footing.

Similar Species: The White-tailed Deer (p. 26), with near-identical prints, prefers denser cover, and it leaves a different high-speed track pattern with a shorter stride. Elk (p. 22) prints are longer and wider. Pronghorn Antelope (p. 28) prints have a wider base.

White-tailed Deer

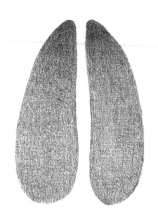

Fore and Hind Prints
Length: 2–3.5 in (5–9 cm)
Width: 1.6–2.5 in (4–6.5 cm)

Straddle
5–10 in (13–25 cm)

Stride
Walking: 10–20 in (25–50 cm)
Galloping: 6–15 ft (1.8–4.5 m)

Size (buck>doe)
Height: 3–3.5 ft (90–110 cm)
Length: to 6.3 ft (1.9 m)

Weight
120–350 lb (55–160 kg)

walking   *gallop group*

WHITE-TAILED DEER
Odocoileus virginianus

The keen hearing of this deer guarantees that it knows about you before you know about it. Frequently, all that we see is its conspicuous white tail in the distance as it gallops away, earning this deer the nickname 'Flagtail.' The White-tailed Deer is not common in this region, but it might be found in small groups at the edges of forests and in brushlands in Idaho. It may be seen around ranches and residential areas.

This deer's prints are heart-shaped and pointed. Its alternating walking track pattern shows the hind prints direct registered or double registered on the fore prints. In snow, or when a deer gallops on soft surfaces, the dewclaws register. This flighty deer gallops in the usual style, leaving hind prints ahead of fore prints, with toes spread wide for steadier, safer footing.

Similar Species: The Mule Deer (p. 24) has nearly identical tracks, but at high speed it stots instead of gallops. Juvenile Elk (p. 22) tracks may be confused with large deer tracks. Pronghorn Antelope (p. 28) prints are also similar.

Pronghorn Antelope

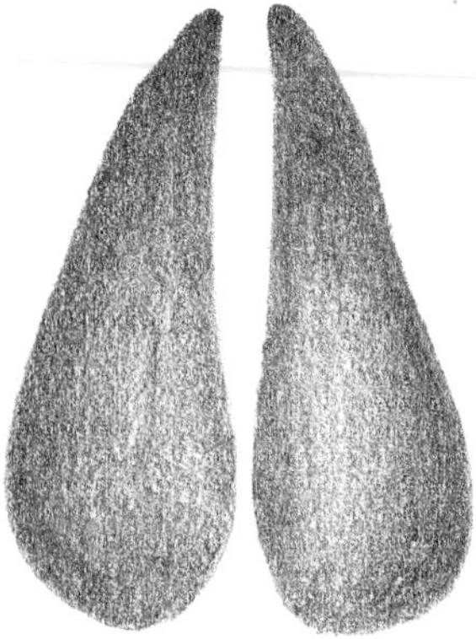

Fore and Hind Prints
Length: 3.3 in (8.5 cm)
Width: 2.5 in (6.5 cm)

Straddle
3.5–9 in (9–23 cm)

Stride
Walking: 8–19 in (20–48 cm)
Galloping: 14 ft (4.3 m) or more

Size (buck>doe)
Height: 3 ft (90 cm)
Length: 3.8–5 ft (1.2–1.5 m)

Weight
34–60 kg (75–130 lb)

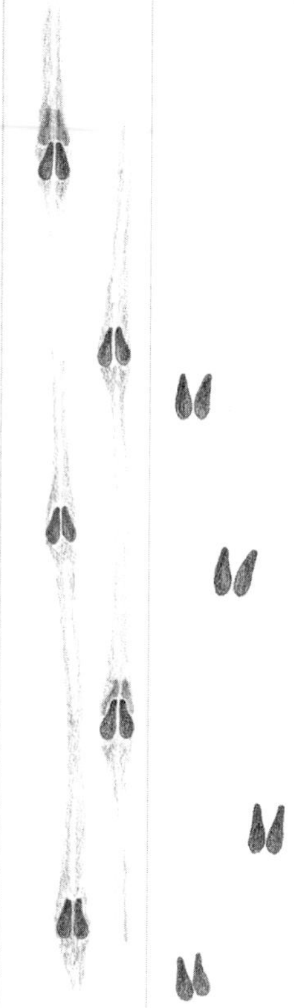

walking

gallop group

PRONGHORN ANTELOPE
Antilocapra americana

The graceful Pronghorn Antelope frequents wide-open grasslands and sagebrush plains across most of this region. Pronghorns gather in groups of up to a dozen animals in summer and as many as 100 in winter, when they prefer to feed in areas where snow has been blown away. Unlike deer, the Pronghorn Antelope—one of North America's fastest animals—appears to run for fun, easily attaining constant speeds of 40 mph (65 km/h) and short spurts up to 60 mph (95 km/h).

A Pronghorn print has a pointed tip and a broad base. This animal does not have dewclaws. The hind prints usually register directly on top of the fore prints, making a tidy alternating track. The Pronghorn tends to drag its feet in snow. When it gallops, the toe tips splay wide. The faster the antelope moves, the greater the distance between gallop groups.

Similar Species: Deer prints (pp. 24–27) show dewclaws, narrower toe bases and shorter strides between gallop groups. Elk (p. 22) prints are similar but larger.

Mountain Goat

Fore and Hind Prints
Length: 2.5–3.5 in (6.5–9 cm)
Width: 2–3.3 in (5–8.5 cm)

Straddle
6.5–12 in (17–30 cm)

Stride
Walking: 10–19 in (25–48 cm)

Size (billy>nanny)
Height: 3–3.5 ft (90–110 cm)
Length: 5–6 ft (1.5–1.8 m)

Weight
100–300 lb (45–140 kg)

walking

MOUNTAIN GOAT

Oreamnos americanus

Spotting the dazzling white coat of a Mountain Goat is truly a wilderness experience. This goat has a preference for high and rugged terrain, such as the rocky slopes of canyons. The Mountain Goat is rarely seen at close range; its tracks in the snow or mud may be your best clue that it is around. This goat, which is native to Idaho, has been introduced into suitable habitat in Oregon, Nevada and Utah.

The goat's squarish print shows long, widely spreading toes. The hard rim and soft middle of the foot help this agile goat clamber over the most unlikely crags at remarkable speeds, but even the Mountain Goat can make a fatal mistake! In deeper snow the feet may leave draglines, and the dewclaws may register. The alternating walking pattern shows a double register, hind over fore.

Similar Species: Deer (pp. 24–27) or Bighorn Sheep (p. 32) prints are smaller, narrower and more pointed. Rugged, remote habitat is usually a good indicator— few deer undertake such rough habitat—but in severe weather the Mountain Goat may come down into deer territory.

Bighorn Sheep

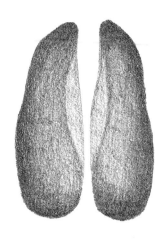

Fore and Hind Prints
Length: 2.5–3.5 in (6.5–9 cm)
Width: 1.8–2.5 in (4.5–6.5 cm)

Straddle
6–12 in (15–30 cm)

Stride
Walking: 14–24 in (35–60 cm)

Size (ram>ewe)
Height: 2.5–3.5 ft (75–110 cm)
Length: 4–6.5 ft (1.2–2 m)

Weight
75–270 lb (34–120 kg)

walking

BIGHORN SHEEP
(Mountain Sheep)
Ovis canadensis

In late fall, the loud crack of two majestic rams head-butting one another can be heard for a great distance. To watch the rut is an awe-inspiring experience, but a rare one. More common in mountain areas than in the Great Basin region, the Bighorn Sheep is listed here as vulnerable and declining. Look for it in areas of suitable rocky habitat.

The squarish print is pointed toward the front. The outer edge of the hoof is hard and the inner part is soft, giving the sheep a good grip on tricky terrain. The neat alternating walking pattern is a direct or double register of the hind atop the fore. When this sheep runs, its toes splay wide. Bighorn Sheep tracks (likely found in groups, because this sheep likes to travel in herds) may lead you to sheep beds—hollows in snow that are used many times and often have an accumulation of dung.

Similar Species: Deer (pp. 24–27) prints are more heart-shaped. The Mountain Goat (p. 30) rarely runs, it enjoys craggier terrain, and its prints are wider at the toe.

Horse

Fore Print
(hind print is slightly smaller)
Length: 4.5–6 in (11–15 cm)
Width: 4.5–5.5 in (11–14 cm)
Straddle
2–7.5 in (5–19 cm)
Stride
Walking: 17–28 in (43–70 cm)
Size
Height: to 6 ft (1.8 m)
Weight
to 1500 lb (680 kg)

walking

HORSE
Equus caballus

   Outdoor adventures on horseback are a popular activity, so you can expect horse tracks to show up almost anywhere. In the Great Basin region there are also a few herds of wild horses.

   Unlike any other animal in this book, the Horse has just one huge toe on each foot. This toe leaves an oval print with a distinctive 'frog' (V-shaped mark) at its base. If a Horse is shod, the horseshoe shows up clearly as a firm wall at the outside of the print. Not all horses are shod, however, so do not expect to see this outer wall on every horse print. A typical, unhurried horse trail shows an alternating walking pattern, with the hind prints registered on or behind the slightly larger fore prints. Horses are capable of a range of speeds—up to a full gallop—but most recreational horseback riders take a more leisurely outlook on life, preferring to walk their horses.

Similar Species: Mules (rarely shod) make smaller tracks.

Black Bear

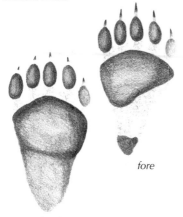

fore

hind

Fore Print
Length: 4–6.3 in (10–16 cm)
Width: 3.8–5.5 in (9.5–14 cm)

Hind Print
Length: 6–7 in (15–18 cm)
Width: 3.5–5.5 in (9–14 cm)

Straddle
9–15 in (23–38 cm)

Stride
Walking: 17–23 in (43–58 cm)

Size (male>female)
Height: 3–3.5 ft (90–110 cm)
Length: 5–6 ft (1.5–1.8 m)

Weight
200–600 lb (90–270 kg)

walking

BLACK BEAR
Ursus americanus

Habitat loss has severely decreased this country's Black Bear numbers, but there are still infrequent sightings in forested areas around the fringe of the Great Basin region. Finding fresh bear tracks can be a thrill, but take care—the bear may be just ahead. Never underestimate the potential power of a surprised bear!

Black Bear prints somewhat resemble small human prints, but they are wider and have claw marks. The small inner toe rarely registers. The forefoot's small heel pad often registers, and the hind print shows a large heel. The bear's slow walk leaves a slightly pigeon-toed double register with the hind print on the fore print. More frequently, at a faster pace, the hind foot oversteps the forefoot. When a bear runs, the two hind feet register in front of the forefeet in an extended cluster. Along well-worn bear paths, look for 'digs' (patches of dug-up earth) and 'bear trees' whose scratched bark shows that this bear climbs.

Similar Species: Black Bear prints cannot be confused with those of any other animal in this region.

Domestic Dog

fore

hind

Fore Print (hind print is smaller)
Length: 1–5.5 in (2.5–14 cm)
Width: 1–5 in (2.5–13 cm)
Straddle
1.5–8 in (3.8–20 cm)
Stride
Walking: 3–32 in (7.5–80 cm)
Loping to Galloping: to 9 ft (2.7 m)
Size
Very variable
Weight
Very variable

walking

loping to galloping

DOMESTIC DOG
Canis familiaris

Dogs come in many shapes and sizes, from the tiny Chihuahua with its dainty feet to the robust and powerful Great Dane. Consequently, Domestic Dog tracks vary enormously. Dog ownership is high in many residential and rural areas, and where dogs are walked or allowed to roam free their tracks are left scattered about, especially if there is wet sand, mud or snow.

The forefeet of the Domestic Dog, which are much larger than the hind feet and support more of the animal's weight, leave the clearest tracks. When a dog walks, the hind prints usually register ahead of or beside the fore prints. As the dog moves faster, it trots and then lopes before it gallops. In a trot or lope pattern the prints alternate fore-hind-fore-hind, whereas a gallop group shows (from back to front) fore-fore-hind-hind.

Similar Species: Keep in mind that dog prints are usually found close to human tracks or activity. Fox (pp. 42–47) prints may be confused with small dog prints, as might Coyote (p. 40) prints, which are usually more oval.

Coyote

fore

hind

**Fore Print
(hind print is slightly smaller)**
Length: 2.4–3.2 in (6–8 cm)
Width: 1.6–2.4 in (4–6 cm)

Straddle
4–7 in (10–18 cm)

Stride
Walking: 8–16 in (20–40 cm)
Trotting: 17–23 in (43–58 cm)
Galloping/Leaping:
    2.5–10 ft (0.8–3 m)

Size (female is slightly smaller)
Height: 23–26 in (58–65 cm)
Length: 32–40 in (80–100 cm)

Weight
20–50 lb (9–23 kg)

*walking
or trotting*

*gallop
group*

COYOTE
(Brush Wolf, Prairie Wolf)
Canis latrans

This widespread, adaptable canine prefers open grasslands or woodlands. On its own, with a mate or in a family pack, it hunts rodents and larger prey. Sometimes a Coyote joins a Badger (p. 68) in hunting for ground squirrels (p. 90). Do not disturb a Coyote den (usually a wide-mouthed tunnel leading into a nesting chamber) or the female will move her pups.

The hind print is slightly smaller than the oval fore print, and its less triangular heel pad rarely registers clearly. The claws of the two outer toes usually do not register. The Coyote typically walks or trots in an alternating pattern; the walk has a wider straddle, and the trotting trail is often very straight. When it gallops, the Coyote's hind feet fall in front of its forefeet; the faster it goes, the straighter the gallop group. The Coyote's tail, which hangs down, leaves a dragline in deep snow.

Similar Species: A Domestic Dog's (p. 38) less oval prints splay more, and its trail is erratic. Foot hairs blur Red Fox (p. 42) prints, which are usually smaller and have a bar across the fore heel pad. Gray Fox (p. 46) and Kit Fox (p. 44) prints are smaller.

41

Red Fox

fore

hind

Fore Print
(hind print is slightly smaller)
Length: 2.1–3 in (5.3–7.5 cm)
Width: 1.6–2.3 in (4–5.8 cm)

Straddle
2–3.5 in (5–9 cm)

Stride
Walking/Trotting:
    12–18 in (30–45 cm)
Side-trotting:
    14–21 in (35–53 cm)

Size (vixen is slightly smaller)
Height: 14 in (35 cm)
Length: 22–25 in (55–65 cm)

Weight
7–15 lb (3.2–7 kg)

trotting   *side-trotting*

RED FOX
Vulpes vulpes

Very adaptable and intelligent, this beautiful and notoriously cunning fox is found scattered throughout the Great Basin, although it is uncommon in some areas. This fox lives in a variety of habitats, from forests to open grasslands.

Abundant foot hair allows just parts of the toes and heel pads to register, with no fine detail. The horizontal or slightly curved bar across the fore heel pad is diagnostic. A trotting Red Fox leaves a distinctive straight alternating trail—the hind print direct registers on the wider fore print. When the fox side-trots, its print pairs show the hind print to one side of the fore print in typical canid fashion. This fox gallops like the Coyote (p. 40). The faster the gallop, the straighter the gallop group.

Similar Species: Other canid prints lack the bar across the fore heel pad. Domestic Dog (p. 38) prints can be of similar size, but with a shorter stride and a less direct trail. Small Coyote prints are similar, but they will have a wider straddle and more-bulbous toe registrations. Gray Fox (p. 46) and Kit Fox (p. 44) tracks are smaller.

Kit Fox

fore

hind

Fore and Hind Prints
Length: 1.1–1.8 in (2.8–4.5 cm)
Width: 1.1–1.5 in (2.8–3.8 cm)

Straddle
2–4 in (5–10 cm)

Stride
Walking/Trotting: 7–10 in (18–25 cm)

Size
Height: 12 in (30 cm)
Length with tail: 24–32 in (60–80 cm)

Weight
3–6 lb (1.4–2.7 kg)

trotting

KIT FOX
Vulpes macrotis

This small, shy fox of the plains and arid regions tends to hide away. Secretive and solitary, its nocturnal activity makes it a rare sight, but its tracks can give it away. Throughout its range, this fox appears to be declining.

The Kit Fox's preference for sandy areas means that its tracks will seldom be very clear, because sand will often fall back into the prints. If you find unclear prints, pay attention to general track characteristics, such as whether the prints are in a typical dog-family trotting pattern. The track's dimensions may help identify which species made them.

Similar Species: The Gray Fox (p. 46) has very similar tracks. The Red Fox (p. 42) has different heel pads (with a bar across them), its prints are generally larger and less clear (because of thick fur), its stride is longer, and its straddle is narrower. Domestic Dog (p. 38) prints of similar size will have a shorter stride and less direct trail. Domestic Cat (p. 52) and Bobcat (p. 50) prints lack claw marks and have larger, less symmetrical heel pads.

Gray Fox

fore

hind

Fore Print
(hind print is slightly smaller)
Length: 1.3–2.1 in (3.3–5.3 cm)
Width: 1.1–1.5 in (2.8–3.8 cm)

Straddle
2–4 in (5–10 cm)

Stride
Walking/Trotting: 7–12 in (18–30 cm)

Size
Height: 14 in (35 cm)
Length: 21–30 in (53–75 cm)

Weight
7–15 lb (3.2–7 kg)

walking

GRAY FOX

*Urocyon
cinereoargenteus*

This small, shy fox is widespread, but it prefers woodlands and chaparral country. The Gray Fox is the only fox that climbs trees, which it does either to seek safety or to forage.

The hind foot does not register as well as the larger forefoot, and its long, semi-retractable claws do not always register. The heel pads are often unclear—they sometimes show up just as small, round dots. This fox has a neat alternating walking track. When it trots, its prints appear in pairs, with the fore print set diagonally behind the hind print. Its gallop group is like the Coyote's (p. 40).

Similar Species: The Red Fox (p. 42) has heel pads with a bar across them, its prints are generally larger and less clear (because of the thick fur), its stride is longer, and its straddle is narrower. The smaller Kit Fox (p. 44) has very similar prints. Coyote and Domestic Dog (p. 38) tracks are much larger. Domestic Cat (p. 52) and Bobcat (p. 50) prints lack claw marks and have larger, less symmetrical heel pads.

Mountain Lion

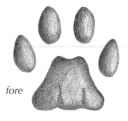

fore

hind

Fore Print
(hind print is slightly smaller)
Length: 3–4.5 in (7.5–11 cm)
Width: 3.3–4.8 in (8.5–12 cm)

Straddle
8–12 in (20–30 cm)

Stride
Walking: 13–32 in (33–80 cm)
Bounding: to 12 ft (3.7 m)

Size
Height: 25–32 in (65–80 cm)
Length: 3.5–5 ft (1.1–1.5 m)

Weight
70–200 lb (32–90 kg)

walking (fast)

MOUNTAIN LION
(Cougar, Puma, Panther)

Puma concolor

Shy, elusive and nocturnal, the Mountain Lion is spread widely but sparsely because of its need for a big home territory. Finding its tracks is usually the best that you can hope for. No longer common in much of its historic range, this native cat is infrequently reported throughout the Great Basin.

Mountain Lion prints tend to be wider than they are long. The retractable claws never register. Thick foot fur enlarges the print in winter and may stop the two lobes on the front of the heel pad from registering clearly. In the walking gait, the hind print direct registers or double registers on the larger fore print. As the pace increases, the hind print tends to fall ahead of the fore print. In snow, the thick, long tail may leave a dragline that can blur print detail. A Mountain Lion seldom gallops, but it is capable of long bounds. Also look for partly buried scat and kills covered for later eating.

Similar Species: Bobcat (p. 50) prints may be confused with juvenile Mountain Lion prints. Coyote (p. 40) tracks are smaller and show claw marks.

Bobcat

fore

hind

**Fore Print
(hind print is slightly smaller)**
Length: 1.8–2.5 in (4.5–6.5 cm)
Width: 1.8–2.5 in (4.5–6.5 cm)

Straddle
4–7 in (10–18 cm)

Stride
Walking: 8–16 in (20–40 cm)
Running: 4–8 ft (1.2–2.4 m)

Size (female is slightly smaller)
Height: 20–22 in (50–55 cm)
Length: 25–30 in (65–75 cm)

Weight
15–35 lb (7–16 kg)

walking

*ambling
to loping*

BOBCAT
(Wildcat)
Lynx rufus

The Bobcat, a stealthy and usually nocturnal hunter, is seldom seen. Very adaptable, it can leave tracks anywhere from wild mountainsides to chaparral and even into residential areas.

A walking Bobcat's hind feet usually register directly on its larger fore prints. As the Bobcat picks up speed its trail becomes an ambling pattern of paired prints, the hind leading the fore. At greater speeds, it leaves four-print groups in a lope pattern. The fore prints, in particular, show asymmetry. The front part of the heel pad has two lobes and the rear part has three. In deep snow the Bobcat's feet leave draglines. Half-buried scat along this cat's meandering trail marks its territory.

Similar Species: A large Domestic Cat (p. 52) will have similar prints but a shorter stride and a narrower straddle, and it will not wander far from home, especially in winter. Mountain Lion (p. 48) prints have similar features but are much larger. Canid (pp. 38–47) prints are narrower than they are long and show claw marks, and the fronts of their footpads have a single lobe.

Domestic Cat

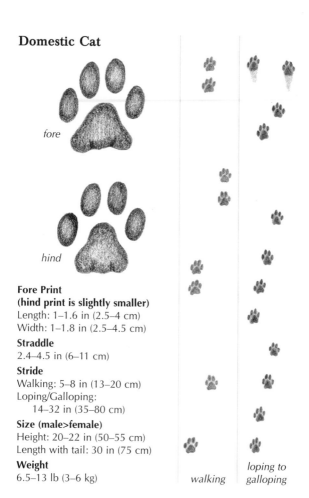

fore

hind

**Fore Print
(hind print is slightly smaller)**
Length: 1–1.6 in (2.5–4 cm)
Width: 1–1.8 in (2.5–4.5 cm)

Straddle
2.4–4.5 in (6–11 cm)

Stride
Walking: 5–8 in (13–20 cm)
Loping/Galloping:
    14–32 in (35–80 cm)

Size (male>female)
Height: 20–22 in (50–55 cm)
Length with tail: 30 in (75 cm)

Weight
6.5–13 lb (3–6 kg)

walking

*loping to
galloping*

DOMESTIC CAT
(House Cat)
Felis catus

The tracks of the familiar and abundant Domestic Cat can show up almost any place where there are people. Abandoned cats may roam farther afield; these 'feral cats' lead a pretty wild and independent existence. Domestic Cats come in many shapes, sizes and colors.

As with all felines, a Domestic Cat's fore print and slightly smaller hind print both show four toe pads. Its retractable claws, kept clean and sharp for catching prey, do not register. Cat prints usually show a slight asymmetry, with one toe leading the others. A Domestic Cat makes a neat alternating walking track pattern, usually in direct register, as you might expect from its fastidious nature. When a cat picks up speed, it leaves clusters of four prints, the hind feet registering in front of the forefeet.

Similar Species: A small Bobcat (p. 50) may leave tracks similar to a very large Domestic Cat's. Canid (pp. 38–47) prints show claw marks.

Ringtail

Fore and Hind Prints
Length: 1–1.4 in (2.5–3.6 cm)
Width: 1–1.4 in (2.5–3.6 cm)

Straddle
3–4 in (7.5–10 cm)

Stride
Walking: 3–6 in (7.5–15 cm)

Size (female is slightly smaller)
Length: 24–32 in (60–80 cm)

Weight
1.5–2.4 lb (0.7–1.1 kg)

walking

RINGTAIL
(Cacomistle, Miner's Cat, Civet Cat)
Bassariscus astutus

This pretty but seldom-seen cousin of the Raccoon (p. 56) may be found in rocky areas and some wooded areas of southern Nevada and southern Utah. The secretive Ringtail is strictly nocturnal and rarely leaves any sign of its passage on the rocky terrain that it frequents. It never goes far from water.

The Ringtail's small, rounded prints show five toes; the partially retractable claws only occasionally register. Just behind the main pad of the fore print, a second pad may be evident. The common walking pattern is an alternating sequence of prints, where the hind foot has registered on or close to the fore print. If you find a Ringtail's trail, it may lead you into rocky talus slopes, up a tree or to the animal's den.

Similar Species: Small mustelids (pp. 60–67) will make similar prints, but the Ringtail has a different gait and habitat, and its fifth toe registers more often. Domestic Cat (p. 52) or Bobcat (p. 50) prints never show five toes or a second pad.

Raccoon

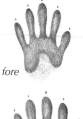

fore

hind

Fore Print
Length: 2–3 in (5–7.5 cm)
Width: 1.8–2.5 in (4.5–6.5 cm)

Hind Print
Length: 2.4–3.8 in (6–9.5 cm)
Width: 2–2.5 in (5–6.5 cm)

Straddle
3.3–6 in (8.5–15 cm)

Stride
Walking: 8–18 in (20–45 cm)
Bounding: 15–25 in (38–65 cm)

Size
(female is slightly smaller)
Length: 24–37 in (60–95 cm)

Weight
11–35 lb (5–16 kg)

walking

bounding group

RACCOON
Procyon lotor

The inquisitive Raccoon, which is found only around the edges of the Great Basin region, is adored by some people for its distinctive face mask, yet disliked for its boundless curiosity—often demonstrated with residential garbage cans. A good place to look for its tracks is near water. The Raccoon likes to rest in trees. It usually dens up for the colder months.

The Raccoon's unusual print, showing five well-formed toes, looks like a human handprint; its small claws appear as dots. Its highly dexterous forefeet rarely leave heel prints, but its hind prints, which are generally much clearer, do show heels. The Raccoon's peculiar walking track pattern shows the left fore print next to the right hind print (or just in front) and vice versa. On the rare occasions when a Raccoon is out in deep snow, it may use a direct-registering walk. The Raccoon occasionally bounds, leaving clusters with the two hind prints in front of the fore prints.

Similar Species: Raccoon prints are normally distinctive, but in snow or mud, River Otter (p. 58) and Yellow-bellied Marmot (p. 88) tracks may look similar.

River Otter

fore

hind

Fore Print
Length: 2.5–3.5 in (6.5–9 cm)
Width: 2–3 in (5–7.5 cm)

Hind Print
Length: 3–4 in (7.5–10 cm)
Width: 2.3–3.3 in (5.8–8.5 cm)

Straddle
4–9 in (10–23 cm)

Stride
Loping: 12–27 in (30–70 cm)

Size
(female is two-thirds the size of male)
Length with tail: 3–4.3 ft (90–130 cm)

Weight
10–25 lb (4.5–11 kg)

loping (fast)

RIVER OTTER
Lontra canadensis

No animal knows how to
have more fun than a River Otter.
If you are lucky enough to watch one at play, you will
not soon forget the experience. Although widespread,
the River Otter is listed as vulnerable and is declining
in the Great Basin region. Well-adapted for the aquatic
environment, this otter lives near water; an otter in
the forest is usually on its way to another waterbody.
Expect to see evidence of an otter's presence along
the waterbodies in its home territory. When the otter
slides down muddy banks, it leaves a trough up to
1 foot (30 cm) wide.

In soft mud, the River Otter's five-toed feet, especially the hind ones, register evidence of webbing.
The inner toes are set slightly apart. If the forefoot's
metacarpal pad registers, it lengthens the print. Very
variable, otter trails usually show the typical mustelid
2×2 loping. However, with faster gaits they show
groups of four and three prints. The thick, heavy
tail often leaves a dragline.

Similar Species: Other mustelid (pp. 60–69) trails lack
conspicuous tail drag and the prints are mostly smaller.

59

Marten

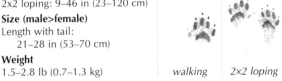

Fore and Hind Prints
Length: 1.8–2.5 in (4.5–6.5 cm)
Width: 1.5–2.8 in (3.8–7 cm)

Straddle
2.5–4 in (6.5–10 cm)

Stride
Walking: 4–9 (10–23 cm)
2x2 loping: 9–46 in (23–120 cm)

Size (male>female)
Length with tail:
    21–28 in (53–70 cm)

Weight
1.5–2.8 lb (0.7–1.3 kg)

walking   *2x2 loping*

MARTEN
(Pine Marten, American Sable)
Martes americana

An aggressive predator, the Marten is rare in the Great Basin region; its numbers are declining throughout much of its natural range, probably because of habitat loss. This handsome weasel-family member prefers wooded areas, especially coniferous and mixed-wood forests.

Marten prints are seldom clear—often just four toes register, and the heel pad is undeveloped. In winter, foot hair often blurs pad detail, especially from the poorly developed palm pads. In the Marten's alternating walking pattern, the hind foot registers on the fore print. In 2×2 loping, the hind prints fall on the fore prints to form slightly angled print pairs in a typical mustelid pattern. Its loping track patterns may appear as three- or four-print clusters (see the River Otter, p. 58). Follow the criss-crossing trails—if a Marten has scrambled up a tree, look for a sitzmark where it has jumped down.

Similar Species: Size and habitat are often key to distinguishing Marten and Mink (p. 62) tracks. Male Mink prints overlap in size with small female Marten prints, but Mink rarely climb trees and (unlike Martens) are usually found near water.

Mink

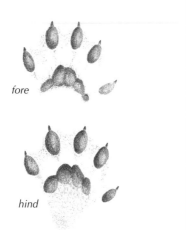

fore

hind

Fore and Hind Prints
Length: 1.3–2 in (3.3–5 cm)
Width: 1.3–1.8 in (3.3–4.5 cm)

Straddle
2.1–3.5 in (5.3–9 cm)

Stride
Walking/2×2 loping: 8–36 in (20–90 cm)

Size (male>female)
Length with tail: 19–28 in (48–70 cm)

Weight
1.5–3.5 lb (0.7–1.6 kg)

2×2 loping

MINK
Mustela vison

The lustrous Mink, which is widespread but uncommon throughout these states, prefers watery habitats surrounded by brush or forest. At home as much on land as in water, this nocturnal hunter can be exciting to track. Like the River Otter (p. 58), the Mink slides in mud and snow, carving out a trough up to 6 inches (15 cm) wide.

The Mink's fore print shows five (perhaps four) toes, with five loosely connected palm pads in an arc, but the hind print shows only four palm pads. The metacarpal pad of the forefoot rarely registers, but the furred heel of the hind foot may register, lengthening the hind print. The Mink prefers the typical mustelid 2×2 loping, making consistently spaced, slightly angled double prints. Its diverse track patterns also include alternating walking; loping with three- and four-print groups (like the River Otter); and bounding (like a rabbit or hare, pp. 74–79).

Similar Species: Small Martens (p. 60) may make similar prints, but without a consistent 2×2 loping gait; they do not live near water. Weasels (pp. 64–67) make similar smaller tracks. Bobcat (p. 50) prints may somewhat resemble four-toed Mink prints.

Long-tailed Weasel

Fore and Hind Prints
Length: 1.1–1.8 in (2.8–4.5 cm)
Width: 0.8–1 in (2–2.5 cm)

Straddle
1.8–2.8 in (4.5–7 cm)

Stride
Bounding: 9.5–43 in (24–110 cm)

Size (male>female)
Length with tail: 12–22 in (30–55 cm)

Weight
3–9.5 oz (85–270 g)

2×2 loping

LONG-TAILED WEASEL
Mustela frenata

Weasels are active year-round hunters that have an avid appetite for rodents. The Long-tailed Weasel is the larger of the two weasels that inhabit the region. Following this nimble creature's tracks can reveal much about its activities. Some weasel trails may lead you up a tree. Weasels sometimes take to water. Tracks are most evident in winter, when a weasel has pursued a rodent over light snow.

The usual weasel gait is a 2×2 lope, leaving a trail of paired prints. The Long-tailed Weasel's typical 2×2 lope shows an irregular stride—sometimes short and sometimes long—with no consistent behavior. Like the Mink (p. 62), this weasel may bound like a rabbit or hare (pp. 74–79).

Similar Species: A large male Short-tailed Weasel's (p. 66) tracks may be the same size as a small female Long-tailed Weasel's. Mink tracks are very similar and only slightly larger.

Short-tailed Weasel

Fore and Hind Prints
Length: 0.8–1.3 in (2–3.3 cm)
Width: 0.5–0.6 in (1.3–1.5 cm)

Straddle
1–2.1 in (2.5–5.3 cm)

Stride
2x2 loping: 9–36 in (23–90 cm)

Size (male>female)
Length with tail: 8–14 in (20–35 cm)

Weight
1.8–6 oz (45–170 g)

2×2 loping

SHORT-TAILED WEASEL
(Ermine, Stoat)
Mustela erminea

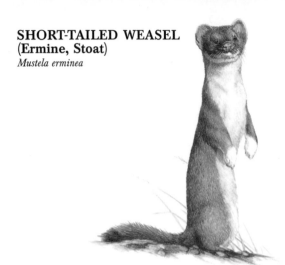

The Short-tailed Weasel is smaller than the Long-tailed Weasel (p. 64). It prefers woodlands and meadows up to higher elevations and does not favor wetlands or dense coniferous forests.

Because of this weasel's light weight, rapid movement and small, hairy feet, pad detail is often unclear, especially in snow. Even with clear tracks, the inner (fifth) toe rarely registers. Although the typical weasel gait is a 2×2 lope that produces a trail of paired prints, this weasel's 2×2 loping tracks may fall in clusters, with alternating short and long strides.

Similar Species: Small female Long-tailed Weasel tracks may be the same size as a large male Short-tailed's.

Badger

fore

hind

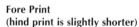

**Fore Print
(hind print is slightly shorter)**
Length: 2.5–3 in (6.5–7.5 cm)
Width: 2.3–2.8 in (5.8–7 cm)

Straddle
4–7 in (10–18 cm)

Stride
Walking: 6–12 in (15–30 cm)

Size
Length: 21–36 in (53–90 cm)

Weight
13–25 lb (6–11 kg)

walking

BADGER
Taxidea taxus

The Badger, with its squat shape and unmistakable face, is most often seen in the open grasslands, but it also ventures into rougher country. It is found throughout the Great Basin region. Thick shoulders and forelegs, coupled with long claws, make it a powerful digger. Look for the Badger's tracks in spring and fall snow—unlike most other mustelids, it likes to den up in a hole for the coldest months of winter.

When a Badger walks, the alternating track pattern shows a double register, with the hind print sometimes falling just behind (or sometimes slightly in front of) the fore print. All five toes on each foot register. A Badger's long claws are evident in the pigeon-toed track that it leaves as it waddles along; the forefoot claws are longer than the hind-foot ones.

Similar Species: On rare occasions, a Porcupine's (p. 82) trail may be similar, but in loose soil, sand or snow its tail and quills leave draglines, and its trail will likely lead up a tree, not to a hole.

Striped Skunk

fore

hind

Fore Print
Length: 1.5–2.2 in (3.8–5.6 cm)
Width: 1–1.5 in (2.5–3.8 cm)

Hind Print
Length: 1.5–2.5 in (3.8–6.5 cm)
Width: 1–1.5 in (2.6–3.8 cm)

Straddle
2.8–4.5 in (7–11 cm)

Stride
Walking/Running:
    2.5–8 in (6.5–20 cm)

Size
Length with tail:
    20–32 in (50–80 cm)

Weight
6–14 lb (2.7–6.5 kg)

walking
(fast)

bounding

STRIPED SKUNK
Mephitis mephitis

   This striking
skunk has a notori-
ous reputation for its vile
smell, and the lingering odor is
often the best sign of its presence. Widespread throughout the region, it prefers lower elevations, in a diversity of habitats. The Striped Skunk dens
up in winter, coming out on warmer days and in spring.
   Both fore and hind feet have five toes. The long
claws on the forefeet often register. The smooth palm
pads and small heel pads leave surprisingly small
prints. The Striped Skunk mostly walks–with such
a potent smell for its defense and those recognizable
black and white stripes, it rarely needs to run. Note that
this skunk's trail rarely shows any consistent pattern,
but an alternating walking pattern may be evident. The
greater a skunk's speed, the more the hind foot
oversteps the fore. If it runs, its trail consists of clumsy,
closely set four-print groups. In snow it drags its feet.

Similar Species: The Western Spotted Skunk (p. 72)
makes smaller prints in a very random pattern. Similar-
sized mustelid (pp. 60–69) tracks will be farther apart
than a skunk's; skunk prints rarely overlap.

Western Spotted Skunk

fore

hind

Fore Print
Length: 1–1.3 in (2.5–3.3 cm)
Width: 0.9–1.1 in (2.3–2.8 cm)

Hind Print
Length: 1.2–1.5 in (3–3.8 cm)
Width: 0.9–1.1 in (2.3–2.8 cm)

Straddle
2–3 in (5–7.5 cm)

Stride
Walking: 1.5–3 in (3.8–7.5 cm)
Jumping: 6–12 in (15–30 cm)

Size
Length: 13–20 in (33–50 cm)

Weight
0.6–2.2 lb (0.3–1 kg)

walking　　*bounding*

WESTERN SPOTTED SKUNK
Spilogale gracilis

This beautifully marked skunk, which is smaller than its striped cousin, is found throughout the Great Basin region. It enjoys diverse habitats such as scrubland, forests and farmland, but it is a rare sight because of its nocturnal habits. It dens up in winter, coming out only on warmer nights.

This skunk leaves a very haphazard trail as it forages for food on the ground. Long claws on the forefeet often register, and the palm and heel may leave defined pad marks. When this skunk runs, which it rarely does, it bounds along leaving groups of four prints, hind in front of fore. It occasionally climbs trees, which it does with ease. It sprays only when truly provoked, so its powerful odor is less frequently detected than that of the Striped Skunk (p. 70).

Similar Species: The Striped Skunk makes larger, less scattered tracks that show a shorter running stride (or jumps); it does not climb trees.

Snowshoe Hare

fore

hind

Fore Print
Length: 2–3 in (5–7.5 cm)
Width: 1.5–2 in (3.8–5 cm)
Hind Print
Length: 4–6 in (10–15 cm)
Width: 2–3.5 in (5–9 cm)
Straddle
6–8 in (15–20 cm)
Stride
Hopping: 0.8–4.3 ft (25–130 cm)
Size
Length: 12–21 in (30–53 cm)
Weight
2–4 lb (0.9–1.8 kg)

hopping

SNOWSHOE HARE
(Varying Hare)
Lepus americanus

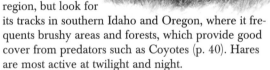

This hare is well known for its color change from summer brown to winter white and for its huge hind feet, which enable it to 'float' on top of snow. This hare is not common in the Great Basin region, but look for its tracks in southern Idaho and Oregon, where it frequents brushy areas and forests, which provide good cover from predators such as Coyotes (p. 40). Hares are most active at twilight and night.

The Snowshoe Hare's most common track pattern is a hopping one that shows triangular four-print groups; they can be quite long if the hare was moving quickly. In winter, heavy fur on the hind feet (much larger than the forefeet) thickens the toes, which can splay out to further distribute the hare's weight on snow. Hares make well-worn runways that are often used as escape runs. You may encounter a resting hare, because hares do not live in burrows. Twigs and stems neatly cut at a 45° angle also indicate this hare's presence.

Similar Species: The Mountain Cottontail (p. 78) has much smaller prints. The Black-tailed Jackrabbit (p. 76) splays its hind toes less.

Black-tailed Jackrabbit

fore

hind

Fore Print
Length: 1.5–3 in (3.8–7.5 cm)
Width: 1.3–1.7 in (3.3–4.3 cm)

Hind Print
Length: 2.5–4 in (6.5–10 cm)
Length with heel: to 6 in (15 cm)
Width: 1.5–2.5 in (3.8–6.5 cm)

Straddle
4–7 in (10–18 cm)

Stride
Hopping: 5–10 ft (1.5–3 m)

Size
Length: 18–25 in (45–65 cm)

Weight
4–8 lb (1.8–3.6 kg)

hopping

BLACK-TAILED JACKRABBIT
Lepus californicus

This athletic hare, which is native to the Great Basin region, frequents open and agricultural areas, and it is sometimes found at higher elevations and in arid zones. Its nocturnal and solitary habits and its wariness of predators result in it being infrequently seen.

Both fore and hind prints show four toes. The hind foot may register a long heel when the hare walks slowly. When it hops, this hare creates print groups in a triangular pattern; as it speeds up, these print groups spread out considerably. Following a hare's trail could lead you to its 'form'—a depression where it rests—or an urgent zigzag pattern that indicates where the hare fled from danger. With its strong hind legs, it is capable of leaping up to 20 feet (6 m) and running up to 35 mph (55 km/h) to avoid pursuers.

Similar Species: The Snowshoe Hare (p. 74) splays its hind feet more. The Mountain Cottontail (p. 78) makes a much smaller print cluster, and it has a shorter stride. Coyote (p. 40) prints resemble heel-less jackrabbit prints, but the pattern will be very different.

Mountain Cottontail

fore

hind

Fore Print
Length: 1–1.5 in (2.5–3.8 cm)
Width: 0.8–1.3 in (2–3.3 cm)

Hind Print
Length: 3–3.5 in (7.5–9 cm)
Width: 1–1.5 in (2.5–3.8 cm)

Straddle
4–5 in (10–13 cm)

Stride
Hopping: 0.6–3 ft (18–90 cm)

Size
Length: 12–17 in (30–43 cm)

Weight
1.3–3 lb (0.6–1.4 kg)

hopping

MOUNTAIN COTTONTAIL
(Nuttall's Cottontail)
Sylvilagus nuttallii

This abundant rabbit
is found throughout
the Great Basin
region. Preferring
brushy areas in
grasslands and cul-
tivated areas, it might
be found in dense vege-
tation, hiding from predators such as the Bobcat (p. 50)
and the Coyote (p. 40). Largely
nocturnal, the Mountain Cottontail might
be seen at dawn or dusk and on darker days.

As with other rabbits and hares, this rabbit's most
common track pattern is a triangular grouping of four
prints, with the larger hind prints (which can appear
pointed) in front of the fore prints (which may overlap).
The hair on the toes will prevent any pad detail from
registering, however. If you track this rabbit, you could
be startled if it flies out from its 'form' (a depression
in the snow or ground in which it rests).

Similar Species: The Desert Cottontail (*S. audubonii*)
makes similar prints. The Snowshoe Hare (p. 74) has
larger prints, especially the hind ones. The Black-tailed
Jackrabbit (p. 76) leaves much larger print clusters, and
it has longer strides. The Pygmy Rabbit (*Brachylagus
idahoensis*) makes slightly smaller prints.

Pika

fore

hind

Fore Print
Length: 0.8 in (2 cm)
Width: 0.6 in (1.5 cm)

Hind Print
Length: 1–1.2 in (2.5–3 cm)
Width: 0.6–0.8 in (1.5–2 cm)

Straddle
2.5–3.5 in (6.5–9 cm)

Stride
Walking/Bounding: 4–10 in (10–25 cm)

Size
Length: 6.5–8.5 in (17–22 cm)

Weight
4–6 oz (110–170 g)

bounding

PIKA
(Cony, Rock Rabbit)
Ochotona princeps

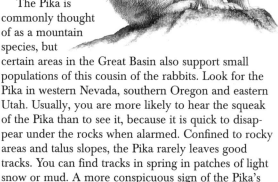

The Pika is commonly thought of as a mountain species, but certain areas in the Great Basin also support small populations of this cousin of the rabbits. Look for the Pika in western Nevada, southern Oregon and eastern Utah. Usually, you are more likely to hear the squeak of the Pika than to see it, because it is quick to disappear under the rocks when alarmed. Confined to rocky areas and talus slopes, the Pika rarely leaves good tracks. You can find tracks in spring in patches of light snow or mud. A more conspicuous sign of the Pika's presence is its little hay piles set to dry in the sun for the winter ahead.

The fore print usually shows five toes unless the fifth toe does not register, but the hind print shows only four. The prints may appear in an erratic alternating pattern or in three- and four-print bounding groups.

Similar Species: Because the Pika is found in rocky habitat, its prints are rarely confused with rabbit or hare (pp. 74–79) prints, which are usually larger, even though track patterns may be similar.

Porcupine

fore

hind

Fore Print
Length: 2.3–3.3 in (5.8–8.5 cm)
Width: 1.3–1.9 in (3.3–4.8 cm)

Hind Print
Length: 2.8–4 in (7–10 cm)
Width: 1.5–2 in (3.8–5 cm)

Straddle
5.5–9 in (14–23 cm)

Stride
Walking: 5–10 in (13–25 cm)

Size
Length with tail: 25–40 in (65–100 cm)

Weight
10–28 lb (4.5–13 kg)

walking

PORCUPINE
Erethizon dorsatum

This notorious rodent rarely runs—its many long quills are a formidable defense. Found throughout the Great Basin region, the Porcupine prefers forests, but it can also be seen in more open areas.

The Porcupine's preferred pigeon-toed waddling gait leaves an alternating walking pattern; the hind print is registered on or slightly in front of the shorter fore print. Look for long claw marks on all prints. The hind print shows five toes, and the fore print shows only four. Clear prints may show the unusual pebbly surface of the solid heel pads, but a Porcupine's tracks are often scratch-marked by its heavy, spiny tail. In deeper snow this squat animal drags its feet, and it may leave a trough with its body. A Porcupine's trail might lead you to a tree, where this animal spends much of its time feeding; if so, look for chewed bark or nipped twigs on the ground.

Similar Species: The Badger (p. 68) makes pigeon-toed prints, but its tracks don't show tail drag, and it doesn't climb trees.

Beaver

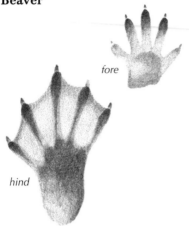

fore

hind

Fore Print
Length: 2.5–4 in (6.5–10 cm)
Width: 2–3.5 in (5–9 cm)
Hind Print
Length: 5–7 in (13–18 cm)
Width: 3.3–5.3 in (8.5–13 cm)
Straddle
6–11 in (15–28 cm)
Stride
Walking: 3–6.5 in (7.5–17 cm)
Size
Length with tail: 3–4 ft (90–120 cm)
Weight
28–75 lb (13–34 kg)

walking

BEAVER
Castor canadensis

Few animals leave as many signs of their presence as the Beaver, North America's largest rodent and a common sight around water. Look for the conspicuous dams and lodges and the stumps of felled trees. Check trunks gnawed clean of bark for marks of the Beaver's huge incisors. Scent mounds marked with castoreum, a strong-smelling yellowish fluid that Beavers produce, also indicate recent activity.

Check the large hind prints for signs of webbing and broad toenails. The nail of the second inner toe usually does not register, and it is rare for all five toes on each foot to do so. Irregular foot placement in the alternating walking gait may produce a direct register or a double register. The Beaver's thick, scaly tail may mar its tracks, as can the branches that it drags about for construction and food. Repeated path use results in well-worn trails.

Similar Species: The Beaver's many signs, including large hind prints, minimize confusion. Muskrat (p. 86) prints are smaller.

Muskrat

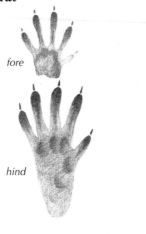

fore

hind

Fore Print
Length: 1.1–1.5 in (2.8–3.8 cm)
Width: 1.1–1.5 in (2.8–3.8 cm)
Hind Print
Length: 1.6–3.2 in (4–8 cm)
Width: 1.5–2.1 in (3.8–5.3 cm)
Straddle
3–5 in (7.5–13 cm)
Stride
Walking: 3–5 in (7.5–13 cm)
Running: to 1 ft (30 cm)
Size
Length with tail: 16–25 in (40–65 cm)
Weight
2–4 lb (0.9–1.8 kg)

walking

MUSKRAT
Ondatra zibethicus

Like the Beaver (p. 84), this rodent is found throughout the Great Basin region, wherever there is water. Beavers are very tolerant of Muskrats and even allow them to live in parts of their lodges. Active all year, the Muskrat leaves plenty of signs. It digs extensive networks of burrows, often undermining riverbanks, so do not be surprised if you suddenly fall into a hidden hole! Also look for small lodges in the water and beds of vegetation on which the Muskrat rests, suns and feeds in summer.

The small fifth (inner) toe of the forefoot rarely registers. Stiff hairs that aid in swimming may create a 'shelf' around the five well-formed toes of the hind print. The common alternating walking pattern shows print pairs that alternate from side to side; the hind print is just behind the fore print or slightly overlaps it. In snow, a Muskrat's feet drag, and its tail leaves a sweeping dragline.

Similar Species: Few animals share this water-loving rodent's habits. The Beaver makes larger tracks and leaves many other signs.

Yellow-bellied Marmot

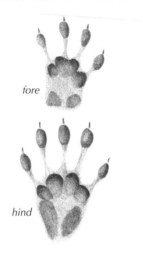

fore

hind

Fore and Hind Prints
Length: 1.5–2.5 in (3.8–6.5 cm)
Width: 1–1.5 in (2.5–3.8 cm)

Straddle
3–5 in (7.5–13 cm)

Stride
Walking: 2–6 in (5–15 cm)
Bounding: 6–14 in (15–35 cm)

Size (male>female)
Length with tail:
    17–28 in (43–70 cm)

Weight
5–10 lb (2.3–4.5 kg)

walking   *bounding*

YELLOW-BELLIED MARMOT
Marmota flaviventris

This endearing large squirrel seems to lead a good life—sleeping all winter and sunbathing on warm rocks in summer. Small colonies of Yellow-bellied Marmots, each with an extensive network of burrows, can be found in most of the Great Basin area, except for the southernmost parts of Nevada and Utah. Marmots are a joy to watch as they play-fight.

The fore print shows four toes and three palm pads; two heel pads may also be evident. The hind print shows five toes, four palm pads and two poorly registering heel pads. In a marmot's usual alternating walking pattern, its hind print falls on top of its fore print. When a marmot runs, it makes groups of four prints, hind ahead of fore. Because this marmot prefers rocky habitats, tracks may be hard to see, but they can be found in spring and fall when light frost or snow cover the ground.

Similar Species: Squirrel (pp. 90, 94–97) prints are very similar but smaller. A small Raccoon's (p. 56) bounding track pattern will be similar, but it will show five-toed fore prints; habitat and behavior are also good indicators.

Great Basin Ground Squirrel

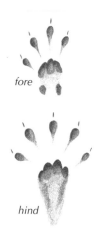

fore

hind

Fore Print
Length: 1–1.3 in (2.5–3.3 cm)
Width: 0.5–1 in (1.3–2.5 cm)

Hind Print
Length: 1.1–1.5 in (2.8–3.8 cm)
Width: 0.8–1.3 in (2–3.3 cm)

Straddle
2.3–3.5 in (5.8–9 cm)

Stride
Bounding: 7–20 in (18–50 cm)

Size
Length with tail: 8–9 in (20–23 cm)

Weight
3.5–5.3 oz (100–150 g)

bounding

GREAT BASIN GROUND SQUIRREL
(Piute Ground Squirrel)

Spermophilus mollis

This small ground squirrel is found only in the Great Basin region. It tends to prefer arid areas dominated by sagebrush and other low shrubs. If you find ground squirrel tracks, the animal itself is probably not far away; it will rarely venture far from the safety of its burrow. Great Basin Ground Squirrels often live together in a loose colony.

This animal's tracks may be evident near its many burrow entrances in mud or in late or early snowfalls; it hibernates in winter. The foreprint rarely shows the small fifth toe, but the two heel pads sometimes show. The larger hind print shows five toes. Both forefeet and hind feet have long claws that frequently register. Ground squirrels are usually seen scurrying around, leaving a typical squirrel track pattern—the hind prints registering ahead of the fore prints, which are usually placed diagonally.

Similar Species: Several other ground squirrels in the region make similar tracks. Chipmunk (p. 92) tracks are smaller. Tree squirrel (pp. 94–97) tracks have a more square-shaped bounding group.

Least Chipmunk

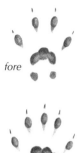

fore

hind

bounding

Fore Print
Length: 0.8–1 in (2–2.5 cm)
Width: 0.4–0.8 in (1–2 cm)

Hind Print
Length: 0.7–1.3 in (1.8–3.3 cm)
Width: 0.5–0.9 in (1.3–2.3 cm)

Straddle
2–3.2 in (5–8 cm)

Stride
Bounding: 7–15 in (18–38 cm)

Size
Length with tail: 7–9 in (18–23 cm)

Weight
1–2.5 oz (28–70 g)

LEAST CHIPMUNK
Tamias minimus

   This delightful chipmunk is found in a variety of
habitats, from dry sagebrush flats to forests, and it is
bold enough to be a popular visitor in campgrounds.
You are more likely to see or hear this rodent, which
is highly active during summer months, than to notice
its tracks. Chipmunks hibernate in winter, but they
occasionally venture out on milder days.
   Chipmunks are so light that their tracks rarely show
fine details. The forefeet each have four toes, and the
hind feet have five. Chipmunks run on their toes, so the
two heel pads of the forefeet seldom register; the hind
feet have no heel pads. Their erratic track patterns, like
those of many of their cousins, show the hind prints
in front of the fore prints. A chipmunk trail often leads
to extensive burrows. Piles of nutshells on rocks are
a further indication of a chipmunk's recent presence.

Similar Species: Several other chipmunk species are
found in this region, and they all make similar tracks;
identifying the species requires a good sighting. Tree
squirrels (pp. 94–97) usually have larger prints and a
wider straddle, and they are more likely to make mid-
winter tracks. Mouse (pp. 106–111) tracks are smaller.

Red Squirrel

fore

hind

Fore Print
Length: 0.8–1.5 in (2–3.8 cm)
Width: 0.5–1 in (1.3–2.5 cm)

Hind Print
Length: 1.5–2.3 in (3.8–5.8 cm)
Width: 0.8–1.3 in (2–3.3 cm)

Straddle
3–4.5 in (7.5–11 cm)

Stride
Bounding: 8–30 in (20–75 cm)

Size
Length with tail:
    9–15 in (23–38 cm)

Weight
2–9 oz (57–260 g)

bounding

*bounding
(deep snow)*

RED SQUIRREL
(Pine Squirrel, Chickaree)
Tamiasciurus hudsonicus

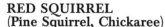

When you enter a Red Squirrel's territory, the inhabitant greets you with a loud, chattering call. Another obvious sign of this forest dweller, which is found in forested areas along the northern and eastern edges of the Great Basin region, is its large middens—piles of cone scales and cores left beneath trees—that indicate favorite feeding sites.

Active all year in its small territory, a Red Squirrel will leave an abundance of trails that lead from tree to tree or down a burrow. This energetic animal mostly bounds, leaving four-print groups, the hind prints in front of the fore prints (which are often side by side). Four toes show on each fore print, and five on each hind print. The heels often do not register when a squirrel moves quickly. In loose soil, sand or snow, the prints merge to form pairs of diamond-shaped tracks.

Similar Species: Chipmunk (p. 92) and Northern Flying Squirrel (p. 96) tracks are in a similar pattern, but they are smaller and have a narrower straddle.

Northern Flying Squirrel

fore

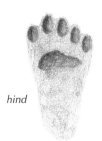

hind

Fore Print
Length: 0.5–0.8 in (1.3–2 cm)
Width: 0.5 in (1.3 cm)

Hind Print
Length: 1.3–1.8 in (3.3–4.5 cm)
Width: 0.8 in (2 cm)

Straddle
3–3.8 in (7.5–9.5 cm)

Stride
Bounding: 11–30 in (28–75 cm)

Size
Length with tail: 9–12 in (23–30 cm)

Weight
4–6.5 oz (110–180 g)

sitzmark into bounding

NORTHERN FLYING SQUIRREL

Glaucomys sabrinus

This soft-furred brown acrobat can be found in coniferous and mixed forests around the edges of this region. It prefers widely spaced forests, where it can use the membranous flaps of skin between its forelegs and hind legs to glide from tree to tree by night. Up to 10 Northern Flying Squirrels den up together in a tree cavity for warmth in winter.

Because it glides, this squirrel leaves fewer tracks than most other squirrels do. Evidence is scarce in summer, but in winter you may find a sitzmark (the distinctive pattern that it made where it landed in the snow) and a short bounding trail, which it made as it rushed off to the nearest tree or to do some quick foraging. The bounding track pattern is typical of squirrels and other rodents, but usually the hind feet register only slightly in front of the forefeet, and often all four feet register in a row.

Similar Species: The Red Squirrel (p. 94) usually makes larger prints and does not leave a sitzmark, but unclear tracks can look the same. Chipmunk (p. 92) prints are slightly smaller, with a narrower straddle.

Bushy-tailed Woodrat

fore

hind

Fore Print
Length: 0.6–0.8 in (1.5–2 cm)
Width: 0.4–0.5 in (1–1.3 cm)

Hind Print
Length: 1–1.5 in (2.5–3.8 cm)
Width: 0.6–0.8 in (1.5–2 cm)

Straddle
2.3–2.8 in (5.8–7 cm)

Stride
Walking: 1.8–3 in (4.5–7.5 cm)
Bounding: 5–8 in (13–20 cm)

Size
Length with tail: 11–19 in (28–48 cm)

Weight
7–21 oz (200–600 g)

walking

BUSHY-TAILED WOODRAT
(Packrat)
Neotoma cinerea

The Bushy-tailed Woodrat is found throughout the Great Basin region, except for southern Nevada. This woodrat thrives in rugged terrain and forests. Tracking one of these nocturnal rodents can be very rewarding: the trail might lead you to its distinctive, massed nest, which can be 5 feet (1.5 m) across and may be in an abandoned building. This animal is a curious hoarder that brings home all manner of objects, thereby serving as a selective wilderness garbage collector.

Four toes show on the fore print and five on the hind. The short claws rarely register. A woodrat often walks in an alternating fashion, with the hind print direct registering on the fore print. This woodrat frequently bounds as well, leaving a pattern of four prints, with the larger hind print in front of the diagonally placed fore prints. The stride tends to be short relative to the size of the prints.

Similar Species: Indistinct Red Squirrel (p. 94) prints are similar but usually larger. The Norway Rat (p. 100) has similar prints but is usually found close to human activity. Marmot (p. 88) prints are similar but much larger.

Norway Rat

fore

hind

Fore Print
Length: 0.7–0.8 in (1.8–2 cm)
Width: 0.5–0.7 in (1.3–1.8 cm)

Hind Print
Length: 1–1.3 in (2.5–3.3 cm)
Width: 0.8–1 in (2–2.5 cm)

Straddle
2–3 in (5–7.5 cm)

Stride
Walking: 1.5–3.5 in (3.8–9 cm)
Bounding: 9–20 in (23–50 cm)

Size
Length with tail: 13–19 in (33–48 cm)

Weight
7–18 oz (200–510 g)

walking

NORWAY RAT
(Brown Rat)
Rattus norvegicus

Active both day and night, this despised rat is widespread almost anywhere that humans have decided to build their homes. Not entirely dependent on people, it may live in the wild as well.

The fore print shows four toes, and the hind print shows five. When it bounds, this colonial rat leaves four-print groups, with the hind prints in front of the diagonally placed fore prints. Sometimes one of the hind feet direct registers on a fore print, creating a three-print group. This rat more commonly leaves an alternating walking pattern with the larger hind prints close to or overlapping the fore prints; the hind heel does not show. The tail often leaves a dragline in loose material. Rats live in groups, so you may find many trails together, often leading to their 5-inch (2-cm) wide burrows.

Similar Species: Woodrat (p. 98) tracks may be similar, but woodrats rarely associate with human activity, except in abandoned buildings. Mouse (pp. 106–111) prints are much smaller. Squirrel (pp. 90, 94–97) tracks show distinctive squirrel traits.

Plains Pocket Gopher

fore

hind

Fore Print
Length: 1 in (2.5 cm)
Width: 0.6 in (1.5 cm)
Hind Print
Length: 0.8–1 in (2–2.5 cm)
Width: 0.5 in (1.3 cm)
Straddle
1.5–2 in (3.8–5 cm)
Stride
Walking: 1.3–2 in (3.3–5 cm)
Size (male> female)
Length with tail: 6–9 in (15–23 cm)
Weight
2.8–5 oz (80–140 g)

walking

PLAINS POCKET GOPHER

Geomys bursarius

This seldom-seen rodent is found throughout the Great Basin region, except southern Nevada. It lives in a variety of habitats, from open forests to grasslands. It spends most of its time in burrows, venturing out only to move mud around or to find a mate. Because of its need to dig, the Plains Pocket Gopher prefers soft, moist soils.

By far the best sign of Plains Pocket Gopher activity is the muddy mounds and tunnel cores—they are especially evident in spring. Each mound marks the entrance to a burrow, which is always blocked up with a plug. Search around the mounds to find tracks. Each foot has five toes. Though the forefeet have long, well-developed claws for digging, the prints rarely show this much detail. Pocket gophers usually walk, leaving an alternating track pattern in which the hind prints fall on or slightly behind the fore prints.

Similar Species: Pocket gopher tracks are associated with their distinctive burrows. Ground squirrels (p. 90) leave rounder tracks with four-toed fore prints.

Long-tailed Vole

fore

hind

Fore Print
Length: 0.5 in (1.3 cm)
Width: 0.5 in (1.3 cm)

Hind Print
Length: 0.6 in (1.5 cm)
Width: 0.5–0.8 in (1.3–2 cm)

Straddle
1.3–2 in (3.3–5 cm)

Stride
Walking/Trotting: 0.8 in (2 cm)
Bounding: 2–6 in (5–15 cm)

Size
Length with tail:
    6.2–8.7 in (15–22 cm)

Weight
1–3 oz (30–87 g)

walking

*bounding
(in snow)*

LONG-TAILED VOLE

Microtus longicaudus

Though a positive identification of a vole track can be challenging, one of the most likely candidates in this region is the Long-tailed Vole. This adaptable rodent lives in a wide variety of habitats.

When clear (which is seldom), a vole's fore print shows four toes, and its hind print shows five. A vole's walk and trot both leave a paired alternating track pattern with a hind print occasionally direct registered on a fore print. Voles usually opt for a faster bounding in which the hind prints register on the fore prints to form print pairs. This vole lopes quickly across open areas, creating a three-print track pattern. Voles stay under the snow in winter; when it melts, look for distinctive piles of cut grass from their ground nests. The bark at the bases of shrubs may show tiny teeth marks left by gnawing. In summer, well-used vole paths appear as little runways in the grass.

Similar Species: Other common voles in the region include the Montane Vole (*M. montanus*) and the Sagebrush Vole (*Lemmiscus curtatus*). Mouse (pp. 106–111) bounding tracks show four-print groups.

Great Basin Pocket Mouse

fore

hind

bounding group

Fore Print
Length: 0.3 in (0.8 cm)
Width: 0.3 in (0.8 cm)
Hind Print
Length: 0.6 in (1.5 cm)
Width: 0.4 in (1 cm)
Straddle
1.5 in (3.8 cm)
Stride
Running: 0.8–4.5 in (2–11 cm)
Size
Length with tail: 3.7–4.8 in (9.5–12 cm)
Weight
0.2–0.3 oz (6–9 g)

bounding

GREAT BASIN POCKET MOUSE

Perognathus parvus

Several pocket mice can be found scattered throughout the Great Basin region. The Great Basin Pocket Mouse prefers to forage by night in thinly vegetated rocky or sandy areas. Although it may hibernate during cold spells, in milder areas it remains active all year.

As a pocket mouse bounds along, it leaves four-print groups in which the two fore prints fall closely side by side and the larger, wider-set hind feet overstep them. The trail is seldom clear. Look for the little depressions in which pocket mice take dust baths to rid themselves of ticks and mites.

Similar Species: Other pocket mice in the region have similar tracks, for example, the widespread, closely related and also nocturnal Little Pocket Mouse (*P. longimembris*). Deer Mouse (p. 108) tracks are similar but slightly larger.

Deer Mouse

hind

fore

bounding group

bounding

bounding (in snow)

Fore Print
Length: 0.3–0.4 in (0.8–1 cm)
Width: 0.3–0.4 in (0.8–1 cm)

Hind Print
Length: 0.3–0.5 in (0.8–1.3 cm)
Width: 0.3–0.4 in (0.8–1 cm)

Straddle
1.4–1.8 in (3.6–4.5 cm)

Stride
Bounding: 6–12 in (15–30 cm)

Size
Length with tail:
6–12 in (15–30 cm)

Weight
0.5–1.3 oz (14–35 g)

DEER MOUSE
Peromyscus maniculatus

The highly adaptable Deer Mouse—one of the region's most abundant mammals—lives anywhere from arid valleys all the way up to alpine meadows. It is seldom seen because it is nocturnal. The Deer Mouse may enter buildings in winter, where it will stay active.

In perfect, soft mud, the fore prints each show four toes, three palm pads and two heel pads, and the hind prints show five toes and three palm pads; the heel pads rarely register. Bounding tracks, most noticeable in snow, show the hind prints falling in front of the fore prints. In soft snow the prints may merge to look like larger pairs of prints; tail drag will be evident. A mouse trail may lead up a tree or down into a burrow.

Similar Species: Many less common mice species have near-identical tracks. The House Mouse (*Mus musculus*), with very similar tracks, associates more with humans. Western Jumping Mouse (p. 110) prints show long, thin toes. Voles (p. 104) tend to trot, and they have a much shorter bounding track pattern. A chipmunk's (p. 92) straddle is wider. A shrew's (p. 114) straddle is narrower.

Western Jumping Mouse

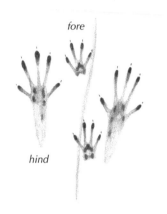

fore

hind

Fore Print
Length: 0.3–0.5 in (0.8–1.3 cm)
Width: 0.3–0.5 in (0.8–1.3 cm)
Hind Print
Length: 0.5–1.3 in (1.3–3.3 cm)
Width: 0.5–0.7 in (1.3–1.8 cm)
Straddle
1.8–1.9 in (4.5–4.8 cm)
Stride
Bounding: 7–18 in (18–45 cm)
In alarm: 3–6 ft (90–180 cm)
Size
Length with tail: 8–10 in (20–25 cm)
Weight
0.6–1.3 oz (17–35 g)

bounding

WESTERN JUMPING MOUSE

Zapus princeps

   Congratulations if you
find and successfully identify the
tracks of the Western Jumping Mouse!
Though it is abundant in the central and
northern parts of the Great Basin region, its preference
for grassy meadows and dense undergrowth, and its
long, deep winter hibernation (about six months!),
make locating tracks very difficult.

   Jumping mouse tracks are distinctive if you do find
them. The two smaller forefeet register between the
long hind prints; the long heels do not always register
and some prints show just the three long middle toes.
The toes on the forefeet may splay so much that the
side toes point backward. When they bound, jumping
mice make short leaps. The tail may leave a dragline
in soft mud or unseasonable snow. Clusters of cut
grass stems, about 5 inches (13 cm) long, found lying
in meadows are a more abundant sign of this rodent.

Similar Species: Deer Mouse (p. 108) tracks may
have the same straddle. Heel-less hind prints may be
mistaken for a vole's (p. 104), or a small bird's (p. 140)
or even an amphibian's (pp. 142–145).

Ord's Kangaroo Rat

fore

hind

slow hop group

Hind Print (fore print is much smaller)
Length: 1.5–1.8 in (3.8–4.5 cm)
Width: 0.5–0.8 in (1.3–2 cm)
Straddle
1.3–2.3 in (3.3–5.8 cm)
Stride
Hopping: 5–24 in (13–60 cm)
Size
Length with tail: 8–14 in (20–36 cm)
Weight
1.5–2.5 oz (43–70 g)

fast hopping

ORD'S KANGAROO RAT
Dipodomys ordii

This small, athletic rodent is capable of big jumps, as its name suggests. This kangaroo rat is common throughout the Great Basin region. In cold winter weather it stays in its burrow, venturing out on milder days.

This nocturnal rodent's preference for drier terrain means that good tracks are hard to find; an abundance of them may be found in sand, but without fine print detail. The best way to identify the track is by its habit. When a kangaroo rat hops slowly, the two small forefeet register between the large hind feet, which show long heel marks, and its long tail leaves a dragline. At higher speed the forefeet do not register, the hind heels appear shorter, and the tail registers infrequently. If you find this kangaroo rat's large nesting mounds, tap your fingers beside a burrow—you may hear thumping in reply.

Similar Species: Other kangaroo rats in this region make identical tracks. Other rats and large mice will register all four feet and their patterns rarely show tail draglines. The Western Jumping Mouse (p. 110) prefers less-arid areas and leaves much smaller tracks.

Vagrant Shrew

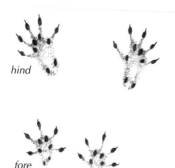

hind

fore

bounding group

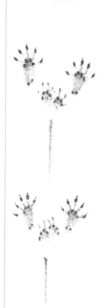

Fore Print
Length: 0.2 in (0.5 cm)
Width: 0.2 in (0.5 cm)
Hind Print
Length: 0.6 in (1.5 cm)
Width: 0.3 in (0.8 cm)
Straddle
0.8–1.3 in (2–3.3 cm)
Stride
Bounding: 1.2–3 in (3–7.5 cm)
Size
Length with tail: 2.3–4.7 in (7–12 cm)
Weight
0.1–0.3 oz (3–9 g)

bounding

VAGRANT SHREW
Sorex vagrans

Though several species of tiny, frenetic shrews are found in this region, the widespread and adaptable Vagrant Shrew is a likely candidate if you find shrew tracks. This small shrew prefers moist woodlands and mixed forests, but it can occasionally be found in open areas if there is adequate moisture.

In its energetic and unending quest for food, a shrew usually leaves a four-print bounding pattern, but it may slow to an alternating walking gait. The individual prints in a group are often indistinct, but in mud or shallow, wet snow you can even count the five toes on each print. In deeper snow a shrew's tail often leaves a dragline. If a shrew tunnels under the snow, it may leave a snow ridge on the surface.

Similar Species: The Dusky Shrew (*S. monticolus*) and Merriam's Shrew (*S. merriami*) both make indistinguishable prints. A mouse's (pp. 106–111) fore prints will show four toes.

Broad-footed Mole

a molehill

some molehills and ridges

Size
Length with tail: 5.2–7.5 in (13–19 cm)
Weight
1.4–2.3 oz (40–65 g)

BROAD-FOOTED MOLE
Scapanus latimanus

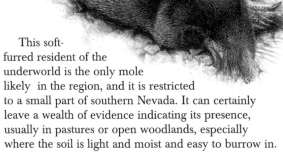

This soft-furred resident of the underworld is the only mole likely in the region, and it is restricted to a small part of southern Nevada. It can certainly leave a wealth of evidence indicating its presence, usually in pastures or open woodlands, especially where the soil is light and moist and easy to burrow in.

Moles, which seldom emerge from their subterranean environment, create an extensive network of burrows through which they forage. These burrows are sometimes marked by ridges on the surface, though most of us are more familiar with the hills that form where the mole gets rid of excess soil from its burrows—moles are frequently considered to be pests when they mess up fine lawns. When rains moisten the soil and bring worms to the surface, it can be entertaining to watch the earth twitch and rise up as the mole satisfies its voracious appetite.

Similar Species: No other moles are common in the Great Basin, but the Plains Pocket Gopher (p. 102) may push up similar piles.

117

BIRDS, AMPHIBIANS & REPTILES

A guide to the animal tracks found in the Great Basin is not complete without some consideration of the birds, amphibians and reptiles found in the region.

Several bird species have been chosen to represent the main types common to this region, but remember that individual bird species are not easily identified by track alone. Bird tracks are often abundant in snow and are clearest in shallow, wet snow. The shores of lakes and streams are very reliable places to find bird tracks—the mud there can hold a clear print for a long time. The sheer number of tracks made by shorebirds and waterfowl can be astonishing. Though some bird species prefer to perch in trees or soar across the sky, it can be entertaining to track birds that frequent the ground. Their tracks can spin around in circles and lead you in all directions. The trail may suddenly end as the bird takes flight, or it might terminate in a pile of feathers, the bird having fallen victim to a predator.

Many amphibians depend on moist environments, so look in the soft mud along the shores of lakes and ponds for their distinctive tracks. You may be able to distinguish frog tracks from toad tracks because frogs and toads generally move differently, but it can be very difficult to identify the species. Reptiles thrive and outnumber the amphibians in drier environments, but they seldom leave good tracks, except in occasional mud or perhaps in sand. Snakes leave distinctive body prints.

Mallard

Print
Length: 2–2.5 in (5–6.5 cm)
Straddle
4 in (10 cm)
Stride
to 4 in (10 cm)
Size
23 in (58 cm)

MALLARD
Anas platyrhynchos

male

female

This dabbling duck—the male a familiar sight with its striking green head—is common in open areas near lakes and ponds. Its webbed prints can often be seen in abundance along the muddy shores of just about any waterbody, including those in urban parks.

The webbed foot of the Mallard has three long toes that all point forward. Though the toes register well, the webbing between the toes does not always show in the print. The Mallard's inward-pointing feet give it a pigeon-toed appearance and perhaps account for its waddling gait, a characteristic for which ducks are known.

Similar Species: Many waterfowl, such as other ducks, as well as the Herring Gull (p. 122) and other gulls, leave similar prints. Large prints are likely made by various species of geese.

Herring Gull

Print
Length: 3.5 in (9 cm)
Straddle
4–6 in (10–15 cm)
Stride
4.5 in (11 cm)
Size
Length: 23–25 in (58–65 cm)

HERRING GULL
Larus argentatus

The Herring Gull, with its long wings and webbed toes, is a strong long-distance flier as well as an excellent swimmer. This bird is increasingly common in the Great Basin region. It congregates in great numbers near waterbodies and garbage dumps.

Gulls leave slightly asymmetrical tracks that show three toes. They have claws that register outside the webbing, and the claw marks are usually attached to the footprint. Most gulls have quite a swagger to their gait, and they leave a trail with the tracks turned strongly inward.

Similar Species: Gull species cannot be reliably identified by track alone, but smaller species have conspicuously smaller tracks. Mallard (p. 120) and other duck tracks are often difficult to distinguish from gull tracks. Geese (various species) make larger prints.

Great Blue Heron

Print
Length: to 6.5 in (17 cm)
Straddle
8 in (20 cm)
Stride
9 in (23 cm)
Size
4.2–4.5 ft (1.3–1.4 m)

GREAT BLUE HERON
Ardea herodias

The refined and graceful image of this large heron symbolizes the precious wetlands in which it patiently hunts for food. Usually still and statuesque as it waits for a meal to swim by, this heron will have cause to walk from time to time, perhaps to find a better hunting location. Look for its large, slender tracks along the banks or mudflats of waterbodies.

Not surprisingly, a bird that lives and hunts with such precision walks in a similar fashion, leaving straight tracks that fall in a nearly straight line. Look for the slender rear toe in the print.

Similar Species: The American Bittern (*Botaurus lentiginosus*) and the Black-crowned Night Heron (*Nyticorax nycticorax*) make similar but smaller tracks. The Sandhill Crane (*Grus canadensis*) leaves prints of similar size, but its smaller rear toes do not register.

Common Snipe

Print
Length: 1.5 in (3.8 cm)

Straddle
to 1.8 in (4.5 cm)

Stride
to 1.3 in (3.3 cm)

Size
11–12 in (28–30 cm)

COMMON SNIPE
Gallinago gallinago

This short-legged character is a resident of marshes and bogs, where its neat prints can often be seen in mud. Snipes are quite secretive when on the ground, and so you may be surprised if one suddenly flushes out from beneath your feet. If there is a Common Snipe in the air, you may hear an eerie whistle if it dives from the sky.

The Common Snipe's neat prints show four toes, including a small rear toe that points inward. The bird's short legs and stocky body give it a very short stride.

Similar Species: Many shorebirds, including the Spotted Sandpiper (p. 128), leave similar tracks.

Spotted Sandpiper

Print
Length: 0.8–1.3 in (2–3.3 cm)

Straddle
to 1.5 in (3.8 cm)

Stride
Erratic

Size
7–8 in (18–20 cm)

SPOTTED SANDPIPER
Actitis macularia

The bobbing tail of the Spotted Sandpiper is a common sight on the shores of lakes, rivers and streams, but you will usually find just one of these territorial birds in any given location. Because of its excellent camouflage, likely the first sign of this bird will be it flying away, its fluttering wings close to the surface of the water.

As it teeters up and down on the shore, a sandpiper leaves trails of three-toed prints. Its fourth toe is very small and faces off to one side at an angle. Sandpiper tracks often have an erratic stride.

Similar Species: All sandpipers and plovers, including the common Killdeer (*Charadrius vociferus*), leave similar tracks, although there is much diversity in size. The Common Snipe (p. 126) makes similar but larger tracks.

Sage Grouse

Print
Length: 2–3 in (5–7.5 cm)
Straddle
2–3 in (5–7.5 cm)
Stride
Walking: 3–6 in (7.5–15 cm)
Size
15–19 in (34–48 cm)

SAGE GROUSE
Centrocercus urophasianus

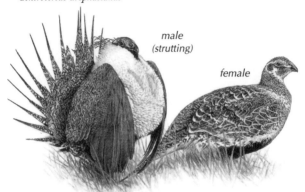

male
(strutting)

female

This ground-dweller tends to spend summers in the foothills and winters on the plains, where its excellent camouflage usually affords it good protection. Although you might find its tracks in mud, snow much improves your chances of finding them. If you follow a Sage Grouse trail quietly, you may be startled when the bird bursts from cover almost beneath your feet.

The three thick front toes leave very clear impressions, but the short rear toe, which is angled off to one side, does not always register well. This bird's neat, straight trail appears to reflect its cautious approach to life on the ground.

Similar Species: The Ruffed Grouse (*Bonasa umbellus*), largely introduced in the Great Basin region, leaves similar tracks.

Great Horned Owl

Strike
Width: to 3 ft (90 cm)
Size
22 in (55 cm)

GREAT HORNED OWL
Bubo virginianus

Often seen resting
quietly in trees by
day, this wide-ranging
owl prefers to hunt at
night. In sand or snow
you might find an untidy
hole, possibly surrounded by
wing and tail-feather imprints.
A well-registered 'strike' can be
quite a sight. The Great Horned
Owl strikes through the snow
with its talons, and the feather
imprints are made as the owl
struggles to take off with possibly
heavy prey. An ungraceful walker,
it prefers to fly away from the scene,
though its tracks may be evident near roadkill.

You might stumble across a strike in snow and guess
that the owl's target could have been a vole (p. 104)
scurrying around underneath the snow. Or you may
be following the surface trail of an animal to find that
it abruptly ends with this mark, where the animal has
been seized.

Similar Species: Several other birds might make
a strike mark; less-rounded, more-distinct feather
imprints could mean that it was made by a hawk
or a Common Raven (*Corvus corax*).

Burrowing Owl

Print
Length: 1.8 in (4.5 cm)

Straddle
2.5 in (6.5 cm)

Stride
1–6 in (2.5–15 cm)

Size
9.5 in (24 cm)

BURROWING OWL
Athene cunicularia

This alert little owl of
the plains and open areas
spends a lot of time on
the ground, bobbing up
and down to look out
for danger. It inhabits
abandoned burrows,
such as those made
by ground squirrels,
often close to bur-
rows still inhabited
by these rodents.
The Burrowing Owl
is an endangered animal
in this region, mainly because
of habitat loss and pesticide use.

If you are investigating rodent burrows in open
areas and you find a burrow entrance with a profusion
of tracks that do not have characteristic rodent features,
you may have found the unusual tracks of the Burrowing Owl. The print of this owl shows two large,
forward-pointing toes with talons, a shorter toe to
the side and a very short fourth one to the rear. The
toes to the side and rear do not register as well as
the front toes, which have more weight on them.

Similar Species: Other owls make similar tracks, but
they are unlikely to be found in Burrowing Owl habitat.

Greater Roadrunner

Print
Length: 3 in (7.5 cm)
Straddle
Standing: 4 in (10 cm)
Stride
(varies with speed) to 12 in (30 cm)
Size
23 in (58 cm)

GREATER ROADRUNNER
Geococcyx californianus

The Greater Roadrunner inhabits the large, open desert region of southern Nevada and Utah. Like the popular cartoon character of the same name, this roadrunner spends most of its time speeding across the plains on its strong legs. With its keen eyesight and tough bill, it hunts and eats insects, small reptiles, rodents and even other birds.

A Greater Roadrunner's foot has two forward-pointing toes and two rearward-pointing toes and leaves a print that is about 3 inches (7.5 cm) in length. The four-pointed star shape of the print is so distinctive that it is unmistakable, even in sand. The Roadrunner hardly ever flies; its preference for staying on the ground and its love for dry areas can result in many tracks for you to discover.

Similar Species: Few other tracks are like those of this ground-loving member of the cuckoo family.

American Crow

Print
Length: 2.5–3 in (6.5–7.5 cm)
Straddle
1.5–3 in (3.8–7.5 cm)
Stride
Walking: 4 in (10 cm)
Size
16 in (40 cm)

AMERICAN CROW
Corvus brachyrhyncos

The black silhouette of the American Crow is a common sight in a variety of habitats. A crow will frequently come down to the ground and contentedly strut around. Its loud *caw* can be heard from quite a distance. Crows can be especially noisy when they are mobbing an owl or hawk.

The American Crow typically leaves an alternating walking track pattern. Its prints show three sturdy toes pointing forward and one toe pointing backward. When a crow is in need of greater speed, perhaps for take-off, it bounds along, leaving irregular pairs of diagonally placed prints with a longer stride between each pair.

Similar Species: Other corvids also spend a lot of time on the ground and make similar tracks: Black-billed Magpie (*Pica hudsonia*) prints are up to 2 inches (5 cm) long. The much larger Common Raven (*C. corax*) leaves tracks up to 4 inches (10 cm) long, and it has a stride to 6 inches (15 cm) in length.

Dark-eyed Junco

Print
Length: to 1.5 in (3.8 cm)
Straddle
1–1.5 in (2.5–3.8 cm)
Stride
Hopping: 1.5–5 in (3.8–13 cm)
Size
5.5–6.5 in (14–17 cm)

DARK-EYED JUNCO
Junco hyemalis

This common small bird typifies the many small hopping birds found in the region. Each foot has three forward-pointing toes and one longer toe at the rear. The best prints are left in mud or light snow; in loose material the toe detail is lost, and the footprints may show some dragging between the hops.

A good place to study this type of prints is near a birdfeeder. Watch the birds scurry around as they pick up fallen seeds, then have a look at the prints left behind. For example, juncos are attracted to small seeds that chickadees (*Poecile* spp.) scatter as they forage for sunflower seeds in the birdfeeder. Also look for tracks under coniferous trees, where juncos feed on fallen seeds in winter.

Similar Species: Toe size may help with identification—larger birds make larger prints—and some birds are seasonal visitors. In fresh, loose snow, junco tracks could be mistaken for mouse (pp. 106–111) tracks, so follow the trail to see if it disappears down a hole or into thin air.

Frogs

fore

hind

Straddle
to 3 in (7.5 cm)

hopping

FROGS

Bullfrog

The best place to look for frog tracks is along the muddy fringes of waterbodies. The smallest frogs include the treefrogs. The most widespread treefrog in the Great Basin is the Pacific Treefrog (*Pseudacris regilla*). It grows to 2 inches (5 cm) in length; it prefers grass and shrubs near water. The Canyon Treefrog (*Hyla arenicolor*) of southern Utah spends its time among rocks and mud along waterbodies, and it can grow to 2.2 inches (5.6 cm) in length. The very adaptable Northern Leopard Frog (*Rana pipiens*), to 5 inches (13 cm) long, is found near clear or brackish water. Unusually large tracks are surely from the robust Bullfrog (*R. catesbeiana*). North America's largest frog, it can grow to 8 inches (20 cm) in length.

A frog's hopping action results in its two small forefeet registering in front of its long-toed hind prints. Frog tracks vary greatly in size, depending on species and age. Toads (p. 144) usually walk, but they may also hop.

Toads

fore

hind

Straddle
to 2.5 in (6.5 cm)

walking

TOADS

Woodhouse's
Toad

The best place to look for toad tracks is, as with frog tracks, undoubtedly along the muddy fringes of water-bodies, but they can occasionally be found in drier areas, for example, as unclear trails in dusty patches of soil. In general, toads walk and frogs (p. 142) hop, but toads are pretty capable hoppers, too, especially when being hassled by overly enthusiastic naturalists. Toads leave rather abstract prints as they walk. The heels of the hind feet do not register. On less-firm surfaces, the toes often leave draglines.

There are fewer toad species than frog species in this region, and they have a very patchy distribution. Woodhouse's Toad (*Bufo woodhousii*) can be found in temporary pools and ditches of the region, except in central Nevada. The Great Plains Toad (*B. cognatus*) and the Red-spotted Toad (*B. punctatus*) both live in grassy areas of southern Nevada and southern Utah. The widespread Great Basin Spadefoot (*Spea intermontana*) is common in forested areas and sagebrush flats. Toads in this region can be up to 5 inches (12 cm) long.

Lizards

fore

hind

Straddle
to 5 in (13 cm)

walking

LIZARDS

Western Fence Lizard

The variety of dry terrain in the Great Basin region is perfectly suited for many species of lizards. Identification of a species merely by a track alone is impossible, especially because the trail is often marred by a dragging belly or tail. Skinks are found in moister areas than other lizards.

If you find lizard tracks, a good candidate is the Western Fence Lizard (*Sceloporus occidentalis*). To over 9 inches (23 cm) in length, it is found throughout the region in both rocky areas and mixed woodlands. Another common lizard is the smaller Desert Horned Lizard (*Phrynosoma platyrhinos*), which grows to just 5.5 inches (14 cm) in length and lives in sandy desert habitats. The large and swift Longnose Leopard Lizard (*Gambelia wislizenii*), named for its leopard-like spots, can grow to over 15 inches (38 cm) in length. It prefers semi-arid habitats throughout the region. The striking Zebra-tailed Lizard (*Callisaurus draconoides*), which can be up to 9 inches (23 cm) in length, lives in flatland areas of central and northwestern Nevada.

These lizards all move very quickly when the need arises, their feet barely touching the ground as they dart for cover. Consequently, their tracks can be hard to make out clearly.

Snakes

SNAKES

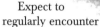

*Gopher
Snake*

Expect to
regularly encounter
many snake species in this region, in a variety of habitats. Because snakes are all long and slender, the tracks of different species appear so similar that accurate identification is next to impossible. In fact, it is very difficult even to tell which way the snake was heading.

Often found throughout the region in many habitats, the harmless and beautiful Gopher Snake (*Pituophis catenifer*) can reach 8.3 feet (2.5 m) in length. The unusual Rubber Boa (*Charina bottae*) lives in grassy and wooded areas; it can be up to 3.5 feet (1.1 m) long. This snake also makes burrows, leaving a wealth of evidence of its presence. Inhabiting water edges, moist meadows and woodlands, the Western Terrestrial Garter Snake (*Thamnophis elegans*) can grow to 3.5 feet (1.1 m) in length. The rattlesnake most likely to be encountered is the Western Rattlesnake (*Crotalus viridis*), which can grow to be 5.3 feet (1.6 m) long. It frequents a variety of habitats, from marshlands to dry woodlands. The Striped Whipsnake (*Masticophis taeniatus*), also common in a variety of habitats, especially shrubby flatland, can be up to 6 feet (1.8 m) long.

TRACK PATTERNS & PRINTS

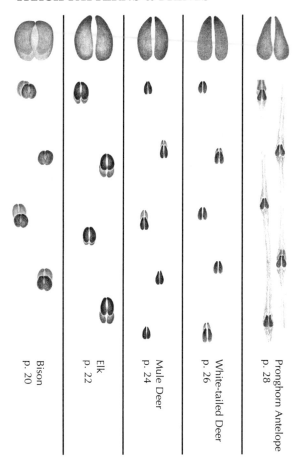

Bison
p. 20

Elk
p. 22

Mule Deer
p. 24

White-tailed Deer
p. 26

Pronghorn Antelope
p. 28

150

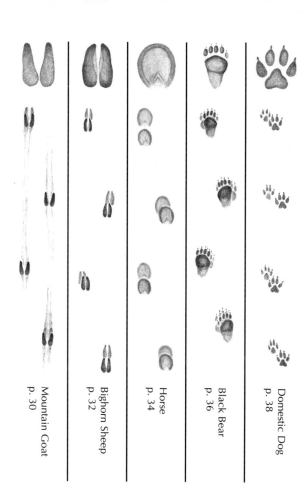

Mountain Goat
p. 30

Bighorn Sheep
p. 32

Horse
p. 34

Black Bear
p. 36

Domestic Dog
p. 38

151

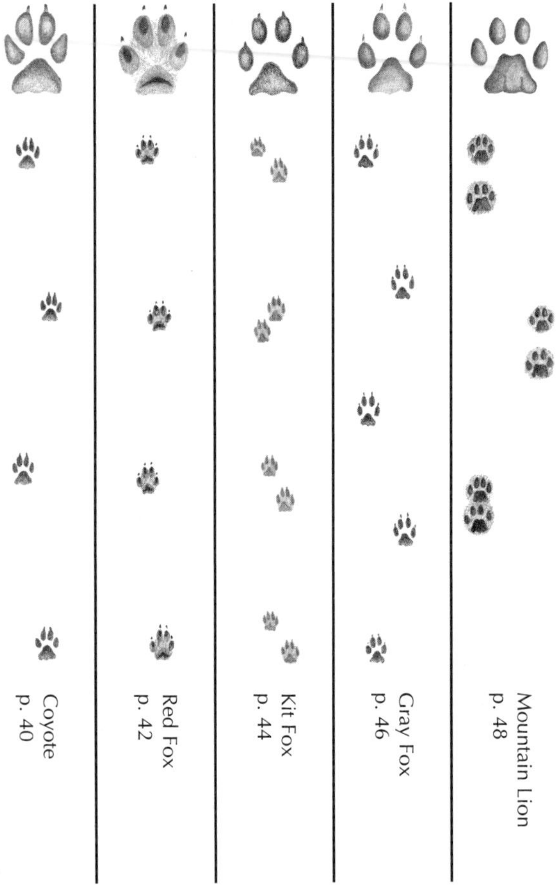

TRACK PATTERNS & PRINTS

Coyote
p. 40

Red Fox
p. 42

Kit Fox
p. 44

Gray Fox
p. 46

Mountain Lion
p. 48

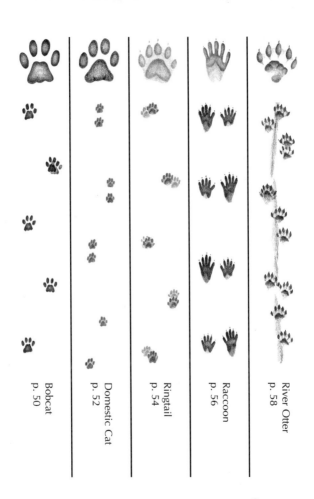

Bobcat
p. 50

Domestic Cat
p. 52

Ringtail
p. 54

Raccoon
p. 56

River Otter
p. 58

153

TRACK PATTERNS & PRINTS

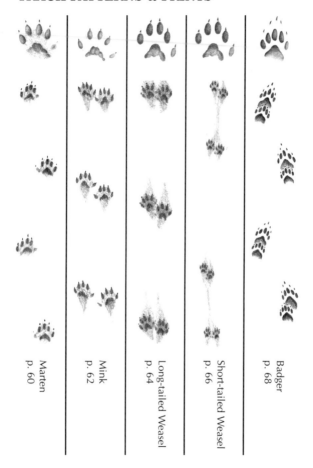

Marten
p. 60

Mink
p. 62

Long-tailed Weasel
p. 64

Short-tailed Weasel
p. 66

Badger
p. 68

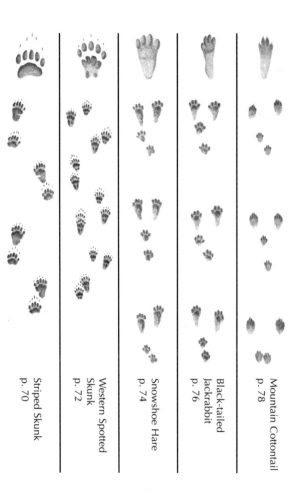

Striped Skunk
p. 70

Western Spotted
Skunk
p. 72

Snowshoe Hare
p. 74

Black-tailed
Jackrabbit
p. 76

Mountain Cottontail
p. 78

TRACK PATTERNS & PRINTS

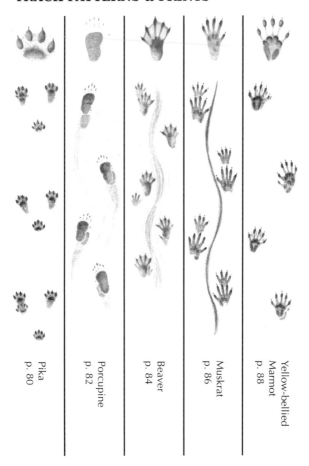

Pika
p. 80

Porcupine
p. 82

Beaver
p. 84

Muskrat
p. 86

Yellow-bellied
Marmot
p. 88

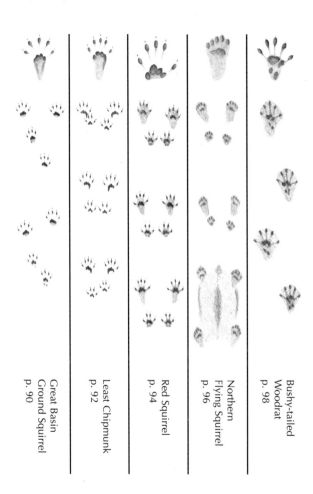

Great Basin
Ground Squirrel
p. 90

Least Chipmunk
p. 92

Red Squirrel
p. 94

Northern
Flying Squirrel
p. 96

Bushy-tailed
Woodrat
p. 98

TRACK PATTERNS & PRINTS

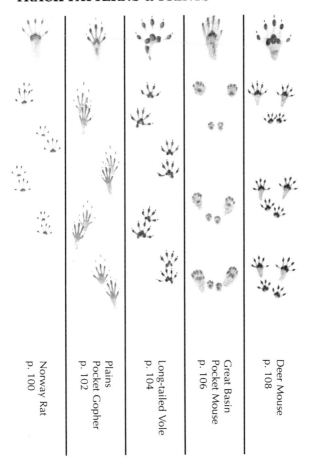

Norway Rat
p. 100

Plains
Pocket Gopher
p. 102

Long-tailed Vole
p. 104

Great Basin
Pocket Mouse
p. 106

Deer Mouse
p. 108

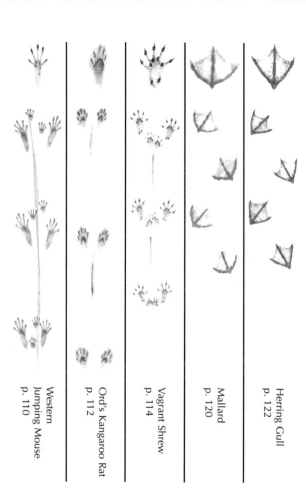

Western
Jumping Mouse
p. 110

Ord's Kangaroo Rat
p. 112

Vagrant Shrew
p. 114

Mallard
p. 120

Herring Gull
p. 122

159

TRACK PATTERNS & PRINTS

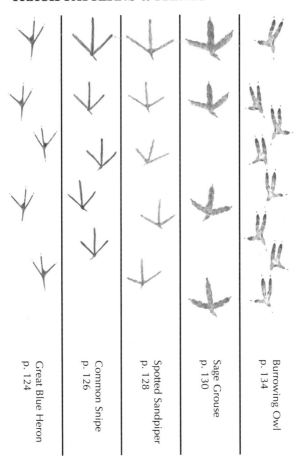

Great Blue Heron
p. 124

Common Snipe
p. 126

Spotted Sandpiper
p. 128

Sage Grouse
p. 130

Burrowing Owl
p. 134

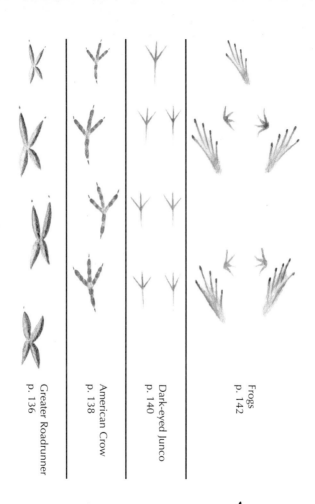

Frogs
p. 142

Dark-eyed Junco
p. 140

American Crow
p. 138

Greater Roadrunner
p. 136

TRACK PATTERNS & PRINTS

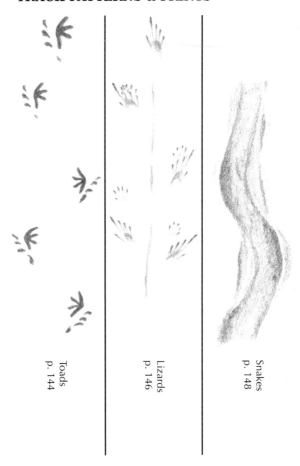

Toads
p. 144

Lizards
p. 146

Snakes
p. 148

HOOFED PRINTS

Mule
Deer

White-tailed
Deer

Bighorn
Sheep

Mountain
Goat

Pronghorn Antelope

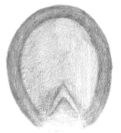

Horse

Elk

Bison

```
inch  cm
0 ─┬─ 0
   │
1 ─┤
   │
2 ─┴─ 5
```

FORE PRINTS

Kit Fox

Gray Fox

Domestic
Cat

Red Fox

Coyote

Bobcat

inch cm
0 ┬ 0
1 ┤
2 ┴ 5

Domestic Dog

Mountain Lion

FORE PRINTS

Western
Spotted
Skunk

Short-tailed
Weasel

Long-tailed
Weasel

Striped
Skunk

Ringtail

Mink

Marten

River Otter

Badger

inch   cm

```
0 ┬ 0


1 ┤


2 ┴ 5
```

HIND PRINTS

Pika

Norway
Rat

Great Basin
Ground
Squirrel

Bushy-tailed
Woodrat

Ord's
Kangaroo
Rat

Northern
Flying
Squirrel

Red
Squirrel

Yellow-bellied
Marmot

Muskrat

Raccoon

Porcupine

Mountain
Cottontail

Snowshoe
Hare

Black-tailed
Jackrabbit

inch cm
0 ⊤ 0

1

2 ⊥ 5

HIND PRINTS

Deer
Mouse

Vagrant
Shrew

Long-tailed
Vole

Great Basin
Pocket
Mouse

inch cm
0 ┬ 0

Northern
Pocket
Gopher

Western
Jumping
Mouse

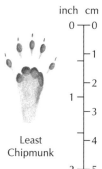
Least
Chipmunk

─ 1

─ 2

1 ─ 3

─ 4

2 ┴ 5

HIND PRINTS

Beaver

Black Bear

inch cm
0 ┬ 0

2 ─

4 ┴ 10

167

BIBLIOGRAPHY

Behler, J.L., and F.W. King. 1979. *Field Guide to North American Reptiles and Amphibians.* National Audubon Society. New York: Alfred A. Knopf.

Brown, R., J. Ferguson, M. Lawrence and D. Lees. 1987. *Tracks and Signs of the Birds of Britain and Europe: An Identification Guide.* London: Christopher Helm.

Burt, W.H. 1976. *A Field Guide to the Mammals.* Boston: Houghton Mifflin Company.

Farrand, J., Jr. 1995. *Familiar Animal Tracks of North America.* National Audubon Society Pocket Guide. New York: Alfred A. Knopf.

Forrest, L.R. 1988. *Field Guide to Tracking Animals in Snow.* Harrisburg: Stackpole Books.

Halfpenny, J. 1986. *A Field Guide to Mammal Tracking in North America.* Boulder: Johnson Publishing Company.

Headstrom, R. 1971. *Identifying Animal Tracks.* Toronto: General Publishing Company.

Murie, O.J. 1974. *A Field Guide to Animal Tracks.* The Peterson Field Guide Series. Boston: Houghton Mifflin Company.

Rezendes, P. 1992. *Tracking and the Art of Seeing: How to Read Animal Tracks and Sign.* Vermont: Camden House Publishing.

Stall, C. 1989. *Animal Tracks of the Rocky Mountains.* Seattle: The Mountaineers.

Stokes, D., and L. Stokes. 1986. *A Guide to Animal Tracking and Behaviour.* Toronto: Little, Brown and Company.

Wassink, J.L. 1993. *Mammals of the Central Rockies.* Missoula: Mountain Press Publishing Company.

Whitaker, J.O., Jr. 1996. *National Audubon Society Field Guide to North American Mammals.* New York: Alfred A. Knopf.

INDEX

Page numbers in **boldface** type refer to the primary (illustrated) treatments of animal species and their tracks.

ABOUT THE AUTHOR

Tamara Eder, equipped from the age of six with a canoe, a dip net and a note pad, grew up with a fascination for nature and the diversity of life. She has a degree in environmental conservation sciences and has photographed and written about the biodiversity in Bermuda, the Galapagos Islands, the Amazon Basin, China, Tibet, Vietnam, Thailand and Malaysia. She is also the author of Lone Pine's Mammal series and Whales series.

Ian Sheldon, an accomplished artist, naturalist and educator, has lived in South Africa, England and Singapore. Living in different countries with different ecosystems encouraged Ian's desire to study mammals, birds and other creatures, and he earned an award from the Zoological Society of London and a degree from Cambridge University. He has also completed a Master's degree in Ecotourism Development. As an artist, Ian is represented by galleries internationally, and he is both a writer and illustrator of many other nature guides, including Lone Pine's *Seashore of Northern and Central California* and the upcoming *Bugs of Northern California*.

Explore the World Outside Your Door

Wildflowers of the Eastern Sierra & Adjoining Mojave Desert & Great Basin

by Laird R. Blackwell

ISBN 1-55105-281-4 • 256 pages • over 700 color photographs • maps

$14.95

A guide to the native and naturalized wildflower species and showy shrubs of the eastern edge of the Sierra Nevada, including the Mojave Desert and Great Basin. The 376 species featured are organized by elevation zone, flower color and petal number.

Birds of Northern California

by David Fix and Andy Bezener

ISBN 1-55105-227-X • 329 pages • 329 color illustrations

$19.95

Learn about 320 species of northern California birds. Descriptions, illustrations and range maps help you identify birds and understand their habits. Perfect for beginner birders and beyond.

Squirrels of the West

by Tamara Hartson

ISBN 1-55105-215-6 • 160 pages • 70 color illustrations • maps

$10.95

This book is a detailed guide to 65 squirrel species found west of the Mississippi. Color illustrations accompany notes on behavior, habitat and ecology of each species.

Order

1-800-518-3541 (Phone)
1-800-548-1169 (Fax)